Instructor's Manual

to accompany

NUMERICAL ANALYSIS
Fourth Edition

By Richard L. Burden and

J. Douglas Faires

prepared by

Richard L. Burden and
J. Douglas Faires

Youngstown State University

PWS-KENT Publishing Company   Boston

PWS-KENT
Publishing Company

20 Park Plaza
Boston, Massachusetts 02116

Copyright © 1989 by PWS-KENT Publishing Company.

All rights reserved. No part of this book may be reproduced
stored in a retrieval system, or transcribed, in any form or
by any means—electronic, mechanical, photocopying, record-
ing, or otherwise—without the prior written permission of
the publisher.

PWS-KENT Publishing Company is a division of Wadsworth, Inc.

ISBN: 0-534-91594-9

Printed in the United States of America.
91 92 93 — 10 9 8 7 6 5 4 3 2

Cover art SPACE & SPACE/NATURE-CLOUD by Susumu Endo, Tokyo,
Japan. A black and white print of an offset lithograph re-
printed with the permission of the artist.

Preface

There are two portions to this Instructors Manual. The first consists of answers to those exercises whose solution is not contained in the text. This portion has been completely redone for the Fourth Edition. Every exercise in the book has been reworked and all the numerical solutions regenerated. In addition, the solutions to the theoretical exercises in the text have been included in this version of the Instructors Manual. Although most of the answers in this portion concern the even-numbered exercises, the solutions for the odd-numbered theoretical exercises that are not given in the answer section of the text are also included. As a consequence, the solution to every exercise is now either in the answer section of the text or in this volume.

The other improvement in the first portion of the Instructors Manual over those of the previous editions concerns the quality of presentation. This volume was set using the TEX document setting system developed by Donald Knuth at Stanford University. The camera-ready copy was then printed an HP laser printer at Youngstown State University . This system provides a level of sophistication that we did not have in our previous editions. We express our appreciation to all those who have been involved in the development of this new tool.

The second part of the Instructors Manual remains basically unchanged from the previous edition. It contains FORTRAN program listings for the algorithms in the text. These programs are written using Structured FORTRAN and executed on the AMDAHL 5868 computer at Youngstown State University. New to the Fourth Edition is a Pascal programing package for the algorithms and the FORTRAN programs listed here parallel the Pascal programs in the new supplement. The programs demonstrate how the results in the examples were obtained and are sufficiently general to apply to the exercises in the text that require computer generated results. Appropriate comments are included to explain the location of each step in the algorithms. Comments are also placed where, for more efficient operation, the program deviates slightly from the algorithm in the text. All programs with the exception of Algorithms 11.6 have been written in single precision mode. but are easily modified to run in double precision. Algorithm 11.6 is listed in double precision because the number of calculations involved in this program requires the increased accuracy.

The authors would like to thank the student assistants who helped to prepare these supplements. They include Brigitte Ramos and Mary Poggione who helped prepare the text of the solutions portion, and Patty Boerio, Francine Byrdy, and Dimitros Chalop, who worked on the algorithms.

Youngstown State University   Richard L. Burden
October 19, 1888   J. Douglas Faires

Table of Contents

1 Mathematical Preliminaries

2 Solutions of Equations of One Variable

3 Interpolation and Polynomial Approximation

4 Numerical Differentiation and Integration

5 Initial–Value Problems for Ordinary Differential Equations

6 Direct Methods for Solving Linear Systems

7 Iterative Techniques in Matrix Algebra

8 Approximation Theory

9 Approximating Eigenvalues

10 Numerical Solutions of Nonlinear Systems of Equations

11 Boundary–Value Problems for Ordinary Differential Equations

12 Numerical Solutions to Partial–Differential Equations

FORTRAN Programs

Chapter 1
Mathematical Preliminaries

Exercise Set 1.1 (PAGE 8)

2. Let $f(x) = x^3 - e^x \sin x$. Since

$$f(1) = -1.287355 < 0 < f(4) = 105.3200,$$

the Intermediate Value Theorem gives the result.

4. $f(0) = f(1) = 0$, so Rolle's Theorem gives the result.

6. Since $f(-1) = 0, f(0) = 0$, and $f(2) = 0$, and f'' exists for all $x > -2$, the Generalized Rolle's Theorem implies that $f''(c) = 0$ for some c in $(-1, 2)$ and hence for some x in $[-1, 3]$.

8. Suppose p and q are in $[a, b]$ with $p \neq q$ and $f(p) = f(q) = 0$. By the Mean Value Theorem, there exists $\xi \in (a, b)$ with

$$f(p) - f(q) = f'(\xi)(p - q)$$

But, $f(p) - f(q) = 0$ and $p \neq q$. So $f'(\xi) = 0$, contradicting the hypothesis.

10. $P_4(x) = e[1 + 2(x - 1) + 3(x - 1)^2 + 10(x - 1)^3/3 + 19(x - 1)^4/6]$.
Since $P_4(1.1) = 3.353408$, $f(1.1) = 3.353485$ and $|R_4(1.1)| \leq 1.10804 \times 10^{-4}$.

12. $P_2(x) = \sin(0) + x \cos(0) + x^2 \frac{\sin(0)}{2} = x$, so $\sin(0.01) \simeq P_2(0.01) = 0.01$.
$R_2(x) = -\frac{\cos(\xi(x))}{6} x^3$ so $|R_2(0.01)| < 1.7 \times 10^{-7}$.
No, in Example 2, $P_2(x) = P_3(x)$, so more accuracy was obtained.

14. $P_3(x) = (x - 1) - \frac{1}{2}(x - 1)^2 + \frac{1}{3}(x - 1)^3$, so $P_3(1.1) = 0.09533333$, whereas $\ln 1.1 = 0.09531018$. $|R_3(1.1)| \leq 2.5 \times 10^{-5}$

16. a) $2, 2 - x, 2 - x + 2x^2, 2 - x + 2x^2 + x^3$
    b) $4, 4 + 6(x - 1), 4 + 6(x - 1) + 5(x - 1)^2, 4 + 6(x - 1) + 5(x - 1)^2 + (x - 1)^3$
    c) $P_m(x) = f(x)$
    d) Since $f^{(m+1)}(x) = 0$ for all x,

$$R_m(x) = \frac{f^{(m+1)}(\xi)}{(m + 1)!}(x - x_0)^{m+1} = 0.$$

17. If f is odd and defined at zero, then since $f(0) = f(-0) = -f(0)$, we must have $f(0) = 0$. If f is even, then f' is odd so $f'(0) = 0$. Also, f'' is even, so f''' is odd and $f'''(0) = 0$. In

a similar manner, $f^{(2n+1)}(0) = 0$ for all positive integers n and the Maclaurin series for f has only even powers.

On the other hand, if the Maclaurin series for f has only even powers, then f is the sum of even functions and hence is an even function.

18. If f is odd, then f' is even. By Exercise 17 its Maclaurin series contains only even powers. Hence, the series for

$$f(x) = \int_0^x f'(t)\, dt$$

contains only odd powers. The converse follows immediately.

20. With $n = 2$, $R_2(1) = \frac{1}{6}e^\xi$ for some ξ in $(0, 1)$. Thus, $|E - R_2(1)| = \frac{1}{6}|1 - e^\xi| \leq \frac{1}{6}(e - 1)$.

22. $f'(x) = 2x - a - b$, $f'(\frac{1}{2}(a + b)) = 0$, and $f(\frac{1}{2}(a + b)) = -(b - a^2)/4$.

24. Using the Mean Value Theorem we have,

$$|g(x_2) - g(x_1)| = |g'(\xi)||x_2 - x_1|$$

for some ξ between x_1 and x_2, so

$$|g(x_2) - g(x_1)| \leq M|x_2 - x_1|.$$

26. a) $f'(0)$ does not exist. Solving $\frac{2}{3}x^{-\frac{1}{3}} = 1$ yields no solution in $[-1, 8]$.
    b) $c = 64/27$.
    c) $f'(0)$ does not exist; $f'(x) \neq 0$ for all $x \neq 0$.
    d) c can be any number in $(0, 1)$.

28. a) Let x_0 be any number in $[a, b]$. Given $\epsilon > 0$, let $\delta = \epsilon/L$.
    If $|x - x_0| < \delta$ and $a \leq x \leq b$, then $|f(x) - f(x_0)| \leq L|x - x_0| < \epsilon$.
    b) See Exercise 24.
    c) $f(x) = x^{1/3}$.

Exercise Set 1.2 (PAGE 18)

2. a) $(149.25, 150.75)$ b) $(895.5, 904.5)$
   c) $(1492.5, 1507.5)$ d) $(89.55, 90.45)$

4. a) 2   b) 3   c) 2   d) 3

6. a) (i) 17/15   (ii) 1.13   (iii) 1.13   (iv) both 3×10^{-3}
   b) (i) 4/15   (ii) 0.266   (iii) 0.266   (iv) both 2.5×10^{-3}

c) (i) 139/660 (ii) 0.211 (iii) 0.210 (iv) 2×10^{-3}, 3×10^{-3}

d) (i) 310/660 (ii) 0.455 (iii) 0.456 (iv) 2×10^{-3}, 1×10^{-4}

8. $(151.55418, 151.70581)$

10. a) $(20/8, 30/7)$

 b) $(20/8, 30/7)$

12. a) $1.37805924 \times 10^{-4}$, $1.37805953 \times 10^{-4}$

 b) $-1.37805953 \times 10^{-4}$, $-1.37805924 \times 10^{-4}$

 c) $144.499985, 144.500015$

 d) $-144.500015, -144.499985$

14. b) first formula: -0.00658

  second formula: -0.0100

The second formula is better since it postpones subtracting nearly equal numbers until the last computation.

16. a) $x = 6.926, y = -12.20$

 b) Subtraction of nearly equal numbers occurs twice in the calculation of y, introducing rounding error. This error is passed on in the calculation of x.

18. $39.375 \leq$ Volume ≤ 86.625, $71.5 \leq$ Surface Area ≤ 119.5

19. Case 1. $d_{k+1} < 5$,

$$\left| \frac{y - fl(y)}{y} \right| = \frac{0.d_{k+1} \ldots \times 10^{n-k}}{0.d_1 \ldots \times 10^n} \leq \frac{0.5 \times 10^{-k}}{0.1} = 5 \times 10^{-k+1}.$$

Case 2. $d_{k+1} > 5$,

$$\left| \frac{y - fl(y)}{y} \right| = \frac{(1 - 0.d_{k+1} \ldots) \times 10^{n-k}}{0.d_1 \ldots \times 10^n} < \frac{(1 - 0.5) \times 10^{-k}}{0.1} = 5 \times 10^{-k+1}.$$

Exercise Set 1.3 (PAGE 27)

2. a) $\frac{1}{1} + \frac{1}{4} \ldots + \frac{1}{100} = 1.53$; $\frac{1}{100} + \frac{1}{81} + \ldots + \frac{1}{1} = 1.54$.

 The actual value is 1.549. Roundoff error occurs much earlier in the first method.

 b) INPUT $N; x_1, x_2, \ldots, x_N$

  OUTPUT SUM

  STEP 1 Set $SUM = 0$

  STEP 2 For $j = 1, \ldots, N$ set $i = N - j + 1$;

     $SUM = SUM + x_i$

STEP 3 OUTPUT (SUM);
STOP.

4. 0.01

6. $O(h)$

8. a) $2,000$ b) $5,000,000$

10. 3

12. INPUT $A; B; C$.
OUTPUT $x_1; x_2$.
STEP 1 If $A = 0$ then
if $B = 0$ then OUTPUT ('NO SOLUTIONS');
STOP
else set $x_1 = -C/B$;
OUTPUT ('ONE SOLUTION',x_1);
STOP.
STEP 2 Set $D = B^2 - 4AC$.
STEP 3 If $D = 0$ then set $x_1 = -B/(2A)$;
OUTPUT ('MULTIPLE ROOTS', x_1);
STOP.
STEP 4 If $D < 0$ then set
$$b = \sqrt{-D}/(2A);$$
$$a = -B/(2A);$$
OUTPUT ('COMPLEX CONJUGATE ROOTS');
set
$$x_1 = a + bi;$$
$$x_2 = a - bi;$$
OUTPUT (x_1, x_2);
STOP.
STEP 5 If $B \geq 0$ then set
$$d = B + \sqrt{D};$$
$$x_1 = -2C/d;$$
$$x_2 = -d/(2A)$$
else set
$$d = -B + \sqrt{D};$$
$$x_1 = d/(2A);$$
$$x_2 = 2C/d.$$
STEP 6 OUTPUT (x_1, x_2);
STOP.

4

14. $x_m = 1 + 1/x_{m-1}$, so $x = 1 + 1/x$. Since $x_m > 0$, we have $x = (1 + \sqrt{5})/2$.

16. 40

18. a)

$$
\begin{aligned}
|F(x) - c_1 L_1 - c_2 L_2| &= |c_1 (F_1(x) - L_1) + c_2 (F_2(x) - L_2)| \\
&\leq |c_1| K_1 |x|^\alpha + |c_2| K_2 |x|^\beta \\
&\leq K |x|^\gamma
\end{aligned}
$$

b)

$$
\begin{aligned}
|G(x) - L_1 - L_2| &= |F_1(c_1 x) + F_2(c_2 x) - L_1 - L_2| \\
&\leq K_1 |c_1 x|^\alpha + K_2 |c_2 x|^\beta \\
&\leq K |x|^\gamma
\end{aligned}
$$

Chapter 2
Solutions of Equations in One Variable

Exercise Set 2.1 (PAGE 36)

2. a) $p_7 = -1.414062$ b) $p_8 = 1.414062$ c) $p_7 = 2.726562$ d) $p_7 = -0.7265625$

4. $p_8 = 3.415039$ using $[3.25, 3.5]$, $p_{10} = 0.5849609$ using $[0, 1]$, and $p_{12} = 3.000061$ using $[1, 3.25]$

6. a) $p_7 = 0.7109375$ b) $p_7 = 1.179687$

8. 14 iterations expected, 1.732

10. 12, 1.379

12. f is symmetric about y-axis; roots are ± 1.9689 and ± 3.1619.

14. Using the interval $[0, 1]$, $h = 0.1617$ so that the depth is $r - h = 0.838$ feet.

16. For $n > 1$,

$$|f(p_n)| = \left(\frac{1}{n}\right)^{10} \le \left(\frac{1}{2}\right)^{10} = \frac{1}{1024} < 10^{-3}$$

so

$$|p - p_n| = \frac{1}{n} < 10^{-3} \Leftrightarrow 1000 < n.$$

Exercise Set 2.2 (PAGE 45)

2. a) (a) $p_4 = 1.10782$ (b) $p_4 = 0.987506$
        (c) $p_4 = 1.12364$ (d) $p_4 = 1.12412$

    b) (d) gives the best answer since $|p_3 - p_4|$ is the smallest.

4. (c), (d); (a) and (b) do not converge.

6. Using $p_0 = 1$ gives $p_{12} = 0.6412053$. Since $|g'(x)| = 2^{-x} \ln 2 \le 0.551$ on $[\frac{1}{3}, 1]$ with $k = 0.551$, $p_0 = 1$ Corollary 2.5 gives a bound of 16 iterations.

8. $p_3 = 1.3231$, $g(x) = \sqrt{1 + \frac{1}{x}}$, $p_0 = 1.5$

10. $p_{14} = 2.92399$, $g(x) = 5/\sqrt{x}$, $p_0 = 2.5$

12. a) $g(x) = \sqrt{\frac{1}{3}e^x}$, $[0,1]$, $p_0 = 0.5$, $p_{14} = 0.910002$

b) $g(x) = \cos x$, $[0,1]$, $p_0 = 0.5$, $p_{28} = 0.7390817$

14. $g(x) = 1/\tan(x) - (1/x) + x$, $p_0 = 4$, $p_4 = 4.493409$

16. a) The proof of existence is unchanged. For uniqueness, suppose p and q are fixed points in $[a, b]$ with $p \neq q$. By the Mean Value Theorem, ξ in (a, b) exists with

$$p - q = g(p) - g(q) = g'(\xi)(p - q) \leq k(p - q) < p - q,$$

giving the same contradiction as in Theorem 2.2.

b) Consider $g(x) = 1 - x^2$ on $[0, 1]$. The function g has the unique fixed point

$$p = \frac{-1 + \sqrt{5}}{2}.$$

With $p_0 = 0.7$, the sequence eventually alternates between 0 and 1.

17. Let $g(x) = x/2 + 1/x$. For $x \neq 0$, $g'(x) = 1/2 - 1/x^2$. If $x > \sqrt{2}$, then $1/x^2 < 1/2$, so $g'(x) > 0$. Also, $g(\sqrt{2}) = \sqrt{2}$.

a) Suppose that $x_0 > \sqrt{2}$. Then

$$x_1 - \sqrt{2} = g(x_0) - g(\sqrt{2}) = g'(\xi)(x_0 - \sqrt{2}),$$

where $\sqrt{2} < \xi < x$.

Thus, $x_1 - \sqrt{2} > 0$ and $x_1 > \sqrt{2}$.

Further,

$$x_1 = \frac{x_0}{2} + \frac{1}{x_0} < \frac{x_0}{2} + \frac{1}{\sqrt{2}} = \frac{x_0 + \sqrt{2}}{2},$$

and $\sqrt{2} < x_1 < x_0$. By an inductive argument,

$$\sqrt{2} < x_{m+1} < x_m < \ldots < x_0.$$

Thus, $\{x_m\}$ is a decreasing sequence which has a lower bound and must converge. Suppose $\lim_{m \to \infty} x_m = p$. Then

$$p = \lim_{m \to \infty} \left(\frac{x_{m-1}}{2} + \frac{1}{x_{m-1}} \right) = \frac{p}{2} + \frac{1}{p}.$$

Thus,

$$p = \frac{p}{2} + \frac{1}{p}$$

and $p = \pm\sqrt{2}$. Since $x_m > \sqrt{2}$ for all m,

$$\lim_{m \to \infty} x_m = \sqrt{2}.$$

b)
$$0 < (x_0 - \sqrt{2})^2 = x_0^2 - 2x_0\sqrt{2} + 2,$$

so
$$2x_0\sqrt{2} < x_0^2 + 2$$

and
$$\sqrt{2} < \frac{x_0}{2} + \frac{1}{x_0} = x_1.$$

c) Case 1: $0 < x_0 < \sqrt{2}$, which implies that $\sqrt{2} < x_1$ by part (b). Thus,

$$0 < x_0 < \sqrt{2} < x_{m+1} < x_m < \ldots < x_1$$

and
$$\lim_{m \to \infty} x_m = \sqrt{2}.$$

Case 2: $x_0 = \sqrt{2}$, which implies that $x_m = \sqrt{2}$ for all m and

$$\lim_{m \to \infty} x_m = \sqrt{2}.$$

Case 3: $x_0 > \sqrt{2}$, which by part (a) implies that

$$\lim_{m \to \infty} x_m = \sqrt{2}.$$

18. a) Let $g(x) = x/2 + A/(2x)$.
Note that $g(\sqrt{A}) = \sqrt{A}$. Also, $g'(x) = 1/2 - A/(2x^2)$ if $x \neq 0$ and $g'(x) > 0$ if $x > \sqrt{A}$.
If $x_0 = \sqrt{A}$, then $x_m = \sqrt{A}$ for all m and

$$\lim_{m \to \infty} x_m = \sqrt{A}.$$

If $x_0 > A$, then

$$x_1 - \sqrt{A} = g(x_0) - g(\sqrt{A}) = g'(\xi)(x_0 - \sqrt{A}) > 0.$$

Further,
$$x_1 = \frac{x_0}{2} + \frac{A}{2x_0} < \frac{x_0}{2} + \frac{A}{2\sqrt{A}} = \frac{1}{2}(x_0 + \sqrt{A}).$$

Thus, $\sqrt{A} < x_1 < x_0$.
Inductively,
$$\sqrt{A} < x_{m+1} < x_m < \ldots < x_0$$

and
$$\lim_{m \to \infty} x_m = \sqrt{A},$$

by an argument similar to that in Exercise 17(a).

If $0 < x_0 < \sqrt{A}$, then

$$0 < (x_0 - \sqrt{A})^2 = x_0^2 - 2x_0\sqrt{A} + A$$

and

$$2x_0\sqrt{A} < x_0^2 + A$$

which leads to

$$\sqrt{A} < \frac{x_0}{2} + \frac{A}{2x_0} = x_1.$$

Thus,

$$0 < x_0 < \sqrt{A} < x_{m+1} < x_m < \ldots < x_1$$

and by the preceding argument,

$$\lim_{m \to \infty} x_m = \sqrt{A}.$$

b) If $x_0 < 0$, then

$$\lim_{m \to \infty} x_m = -\sqrt{A}.$$

20. Let $\varepsilon = (1 - |g'(p)|)/2$. Since g' is continuous at p, there exists $\delta > 0$ such that for $x \in [p - \delta, p + \delta]$,

$$|g'(x) - g'(p)| < \varepsilon.$$

Thus, $|g'(x)| < |g'(p)| + \varepsilon < 1$ for $x \in [p - \delta, p + \delta]$.
By the Mean Value Theorem

$$|g(x) - g(p)| = |g'(c)||x - p| < |x - p|$$

for $x \in [p - \delta, p + \delta]$. Applying Theorem 2.3 completes the problem.

Exercise Set 2.3 (Page 54)

2. a) With $p_0 = 1$ and $p_1 = 4$, $p_{11} = 2.690648$.
  b) With $p_0 = -2$ and $p_1 = 0$, $p_7 = -0.6527037$. With $p_0 = -4$ and $p_1 = -2$,
    $p_{11} = -2.414795$.
  c) With $p_0 = 0$ and $p_1 = 1.570796$, $p_6 = 0.7390850$.
  d) With $p_0 = 0$ and $p_1 = 1.570796$, $p_5 = 0.9643338$.

4. a) 0.257530, 3 iterations with $p_0 = 0.0$
  b) -0.458962, 5 iterations with $p_0 = 0.0$; 0.910008, 3 iterations with $p_0 = 1.0$
  c) 1.829384, 7 iterations with $p_0 = 1.0$

d) 3.161950, 4 iterations with $p_0 = 3.0$; 1.968873, 3 iterations with $p_0 = 1.0$

6. $(0.5897546, 0.3478105)$

8. a) 0.90479, 3 iterations with $p_0 = 1.0$ b) 0.90479, 5 iterations

9. The equation of the tangent line is

$$y - f(p_{m-1}) = f'(p_{m-1})(x - p_{m-1}).$$

To complete the problem, set $y = 0$ and solve for $x = p_m$.

10. The x-intercept of the secant line from $(x_0, f(x_0))$ to $(x_1, f(x_1))$ occurs at x_2.

12. a) 1.7321, 3 iterations   b) 1.7321, 5 iterations, $p_0 = 1$, $p_1 = 2$

14. If $p_0 < 1.75 < p_1 < 2$, the Secant Method will work.

16. Changing STEP 6 in Algorithm 2.4 as follows will give the desired result.

STEP 6 Set $q = f(p)$;

If $q \cdot q_1 < 0$ then

set

$$p_0 = p_1; \quad q_0 = q_1; \quad p_1 = p; \quad q_1 = q.$$

else

set

$$p_1 = p; \quad q_1 = q.$$

18. 0.739085, 5 iterations

20. a) $p_0 = -0.5$ yields $p_2 = -0.4341431$
    b) $p_0 = 0.5$ yields $p_3 = 0.4506567$
       $p_0 = 1.5$ yields $p_3 = 1.7447381$
       $p_0 = 2.5$ yields $p_4 = 2.2383198$
       $p_0 = 3.5$ yields $p_3 = 3.7090412$
    c) $n - 0.5$
    d) $p_0 = 24.5$ yields $p_2 = 24.4998870$

22. 0.842

24. 6.75%, 6.63%

25. 10.3% per year

26. $P_L = 259300$, $c = -0.720674$, $k = 0.047988$
    $P(1980) = 221,470,000$, $P(2000) = 243,379,000$

28. $p_3 = -0.31706178$

30. b) 33.2°

Exercise Set 2.4 (PAGE 66)

2. $p_{10} = -0.5671443$, $p_3 = -0.5671438$

4. $p_4 = -0.1696067$

6. $N > 10^{m/k}$

8. $p_n = 10^{-\alpha^n}$

10. With $m = 2$, $p_{10} = -0.5671417$

12.

	3	4
$m = 1$	$p_{17} = -0.1830455$	$p_{10} = -0.1696068$
$m = 2$	$p_7 = -0.1830000$	$p_{20} = -0.1679994$ (failed)
$m = 3$	$p_6 = -0.1831835$	$p_{20} = -0.1759686$ (failed)
$m = 4$	$p_{20} = -0.1948774$ (failed)	$p_{19} = -0.1869667$ (failed)

14. If f has a zero of multiplicity m at p, then f can be written as

$$f(x) = (x - p)^m q(x),$$

for $x \neq p$, where

$$\lim_{x \to p} q(x) \neq 0.$$

Thus,

$$f'(x) = m(x - p)^{m-1} q(x) + (x - p)^m q'(x)$$

and $f'(p) = 0$. Also,

$$f''(x) = m(m - 1)(x - p)^{m-2} q(x) + 2m(x - p)^{m-1} q'(x) + (x - p)^m q''(x)$$

and $f''(p) = 0$.

In general, for $k \leq m$,

$$f^{(k)}(x) = \sum_{j=0}^{k} \binom{k}{j} \frac{d^j (x - p)^m}{dx^j} q^{(k-j)}(x)$$

$$= \sum_{j=0}^{k} \binom{k}{j} m(m - 1) \cdots (m - j + 1)(x - p)^{m-j} q^{(k-j)}(x).$$

Thus, $f^{(k)}(p) = 0$ for $0 \leq k \leq m - 1$, but

$$f^{(m)}(p) = m! \lim_{x \to p} q(x) \neq 0.$$

If $f(p) = f'(p) = \ldots = f^{(m-1)}(p) = 0$ and $f^{(m)}(p) \neq 0$, consider the $(m-1)$th Taylor polynomial of f expanded about p :

$$f(x) = f(p) + f'(p)(x-p) + \ldots + \frac{f^{(m-1)}(p)(x-p)^{m-1}}{(m-1)!} + \frac{f^{(m)}(\xi)(x-p)^m}{m!}$$

$$= (x-p)^m \frac{f^{(m)}(\xi)}{m!},$$

where ξ is between x and p.

Since $f^{(m)}$ is continuous, let

$$q(x) = \frac{f^{(m)}(\xi(x))}{m!}.$$

Then $f(x) = (x-p)^m q(x)$ and

$$\lim_{x \to p} q(x) = \frac{f^{(m)}(p)}{m!} \neq 0.$$

15. Let

$$g(x) = x - \frac{f(x)}{f'(x)} - \frac{f''(x)}{2f'(x)} \left[\frac{f(x)}{f'(x)} \right]^2$$

Show that if $f(p) = 0$, then

$$g(p) = p, \text{ and } g'(p) = g''(p) = g'''(p) = 0.$$

If $\frac{|e_{m+1}|}{|e_m|^3} = 0.75$ and $|e_0| = 0.5$, then

$$|e_m| = (0.75)^{\frac{3^m-1}{2}} |e_0|^{3^m}$$

To have $|e_m| \leq 10^{-8}$ requires that $m \geq 3$.

16. Let $e_n = p_n - p$. If

$$\lim_{n \to \infty} \frac{|e_{n+1}|}{|e_n|^\alpha} = \lambda > 0,$$

then for sufficiently large values of n, $|e_{n+1}| \simeq \lambda |e_n|^\alpha$.
Thus,

$$|e_n| \simeq \lambda |e_{n-1}|^\alpha \text{ or } |e_{n-1}| \simeq \lambda^{-1/\alpha} |e_n|^{1/\alpha}.$$

Using the hypothesis gives

$$\lambda |e_n|^\alpha \simeq c|e_n| \lambda^{-1/\alpha} |e_n|^{1/\alpha}$$

so

$$|e_n|^\alpha \simeq c\lambda^{-1/(\alpha-1)} |e_n|^{1+1/\alpha}$$

12

Since the powers of $|e_n|$ must agree,

$$\alpha = 1 + 1/\alpha \quad \text{so} \quad \alpha = \frac{1 + \sqrt{5}}{2} \simeq 1.62.$$

17. a) For any $\epsilon > 0$ there exists an integer N such that, for $m \geq N$,

$$\left| \frac{|p_{m+1} - p|}{|p_m - p|} - \lambda \right| < \epsilon.$$

Thus, for $m \geq N$,

$$\lambda - \epsilon < \frac{|p_{m+1} - p|}{|p_m - p|} < \lambda + \epsilon.$$

So

$$(\lambda - \epsilon)|p_m - p| < |p_{m+1} - p| < (\lambda + \epsilon)|p_m - p|,$$

$$(\lambda - \epsilon)^2|p_{m-1} - p| < |p_{m+1} - p| < (\lambda + \epsilon)^2|p_{m-1} - p|,$$

and

$$(\lambda - \epsilon)^{m+1-N}|p_N - p| < |p_{m+1} - p| < (\lambda + \epsilon)^{m+1-N}|p_N - p|.$$

(*i*) If $\lambda < 1$, choose $\epsilon > 0$ satisfying $\lambda + \epsilon < 1$. Then

$$0 \leq \lim_{m \to \infty} |p_{m+1} - p| \leq \lim_{m \to \infty} (\lambda + \epsilon)^{m+1-N}|p_N - p| = 0.$$

(*ii*) If $\lambda > 1$, choose $\epsilon > 0$ satisfying $1 - \epsilon > 1$. Then,

$$0 < \lim_{m \to \infty} (\lambda - \epsilon)^{m+1-N}|p_N - p| \leq \lim_{m \to \infty} |p_{m+1} - p|$$

and $\{p_m\}_{m=0}^{\infty}$ cannot converge to p.

(*iii*) The sequence may converge, for example when $p_m = 1/m$ and $p = 0$.

Or the sequence may diverge, for example when $p_m = m$ and $p = 0$.

b) The proof is similar to a).

Exercise Set 2.5 (PAGE 72)

2. With $p_0^{(0)} = 0.5$, $p_3^{(0)} = 0.6411857$

4. a) 0.257530, 3 iterations

b) $g(x) = (e^x/3)^{1/2}$, 0.910008, 4 iterations

c) 0.739085, 3 iterations

6. a) $\hat{p}_{10} = 0.0\overline{45}$ **b)** $\hat{p}_2 = 0.036324786$

7. a) A positive constant λ exists with

$$\lambda = \lim_{n \to \infty} \frac{|p_{n+1} - p|}{|p_n - p|^\alpha}.$$

Hence

$$\lim_{n \to \infty} |\frac{p_{n+1} - p}{p_n - p}| = \lim_{n \to \infty} \frac{|p_{n+1} - p|}{|p_n - p|^\alpha} \cdot |p_n - p|^\alpha = \lambda \cdot 0 = 0$$

and

$$\lim_{n \to \infty} \frac{p_{n+1} - p}{p_n - p} = 0.$$

b) One example is $p_n = \frac{1}{n^n}$.

8.

$$\frac{|p_{m+1} - p_m|}{|p_m - p|} = \frac{|p_{m+1} - p + p - p_m|}{|p_m - p|} = \left| \frac{p_{m+1} - p}{p_m - p} + 1 \right|.$$

So

$$\lim_{m \to \infty} \frac{|p_{m+1} - p_m|}{|p_m - p|} = \lim_{m \to \infty} \left| \frac{p_{m+1} - p}{p_m - p} + 1 \right| = 1.$$

Exercise Set 2.6 (PAGE 81)

2. a) 4.123106, −4.123106, −2.5 + 1.322876i, −2.5 − 1.322876i

  b) 4.381113, −3.548233, 0.5835597 + 1.494188i, 0.5835597 − 1.494188i

  c) 1.414214i, −1.414214i, −0.5 + 0.8660254i, −0.5 − 0.8660254i

  d) −12.61243, 2.260086, −0.2502369, −0.1987093 + 0.8133672i, −0.1987093 − 0.8133672i

  e) −3.358045, −0.8467426, −1.494349 + 1.744218i, −1.494349 − 1.744218i

  f) 2.069323, 0.8611735, −1.465250 + 0.8116790i, −1.465250 − 0.8116790i

  g) 2.732051, 1.414214, −1.414214, −0.7320508

  h) 3.414214, 3.0000000, 0.5857864

4. a) −2.5 + 1.322875i, −2.5 − 1.322875i, 4.123104, −4.123102

  b) 0.5835596 + 1.494188i, 0.5835596 − 1.494188i, 4.381114, −3.548233

  c) 1.414213i, −1.414213i, −0.4999999 + 0.8660258i, −0.4999999 − 0.8660258i

  d) −0.2502371, −0.1987094 + 0.8133125i, −0.1987094 − 0.8133125i, 2.260085, −12.61242

  e) 0.8467426, −1.494350 + 1.744219i, −1.494350 − 1.744219i, −3.358037

  f) 0.8611736, −1.465248 + 0.8116722i, −1.465248 − 0.8116722i, 2.069322

  g) 1.414213, −0.7320500, 2.732050, −1.414214

  h) 0.5857863, 3.414211, 3.000002

14

5. a) By Theorem 2.6, P has at least one zero x_1, which may be complex. Dividing P by $(x - x_1)$ gives

$$P(x) = (x - x_1)Q_1(x) + R_1(x),$$

where deg $Q_1 = n - 1$ and deg $R_1 <$ deg $(x - x_1) = 1$.
Since R_1 must be constant and $P(x_1) = 0$, it follows that $R_1(x) \equiv 0$ and

$$P(x) = (x - x_1)Q_1(x).$$

Since Q_1 is a polynomial of deg $n - 1$, it must have a zero x_2 (assuming $n > 1$). Thus, Q_1 can be factored into $Q_1(x) = (x - x_2)Q_2(x)$ where deg $Q_2 = n - 2$. Hence,

$$P(x) = (x - x_1)(x - x_2)Q_2(x).$$

This process can be continued up to the degree n of P to obtain numbers $x_1, x_2, \ldots, x_n$ and a polynomial $Q_n(x)$ of degree zero for which

$$P(x) = (x - x_1)(x - x_2)\cdots(x - x_n)Q_n(x).$$

Expanding the product $(x - x_1)(x - x_2)\cdots(x - x_n)$ gives

$$(x - x_1)(x - x_2)\cdots(x - x_n) = x^n + \{\text{lesser degree terms}\}.$$

The constant polynomial $Q_n(x)$ must be the constant a_n. Hence,

$$P(x) = a_n(x - x_1)(x - x_2)\cdots(x - x_n).$$

Note that this proves that P has exactly n roots since the division process cannot be continued.
Collecting multiple roots, and relabeling if necessary, leads to the existence of distinct roots $x_1, \ldots, x_k$ and multiplicities $m_1, \ldots, m_k$, where $\sum_{i=1}^{k} m_i = n$. As a consequence,

$$P(x) = a_n(x - x_1)^{m_1}(x - x_2)^{m_2}\cdots(x - x_k)^{m_k}.$$

b) If P and Q are polynomials of degree at most n, then $D(x) = P(x) - Q(x)$ is a polynomial of degree at most n. Since $P(x_i) = Q(x_i)$ for the distinct numbers $x_1, \ldots, x_k$, it follows that D has the k distinct zeros $x_1, \ldots, x_k$. Thus, D can be written as

$$D(x) = (x - x_1)(x - x_2)\cdots(x - x_k)D_1(x),$$

where D_1 is a polynomial. However,

$$\deg D \geq \deg(x - x_1)(x - x_2)\cdots(x - x_k) = k > n = \deg D.$$

15

The only alternative to this obvious contradiction is for $D(x)$ to be the zero polynomial. Thus, $D(x) \equiv 0$ and $Q(x) \equiv P(x)$.

6. Note that

$$(x - a - bi)(x - c - di) = x^2 - (a + c + bi + di)x + ac - bd + (ad + bc)i,$$

which can only be real if $b = -d$ and $a = c$.

7. Let $x_1 > x_0$. Then $x_1 - x_0 > 0$, and since $b_i > 0$ for $i = 1, 2, \ldots n, x_1 > 0$, we have

$$b_n x_1^{n-1} + \ldots + b_2 x_1 + b_1 > 0.$$

Thus,

$$P(x_1) = (x_1 - x_0)(b_n x_1^{n-1} + \cdots + b_2 x_1 + b_1) + P(x_0) > 0$$

and P cannot have any real zeros greater than x_0.

8. Convergence to the root 0.27. We need p_0 closer to 0.29, since $f'(0.28\overline{3}) = 0$.

10. Since the volume $V = 1000 = \pi r^2 h$, $h = 1000/(\pi r^2)$. The amount of material $M(r)$ is given by

$$M(r) = 2\pi(r + 0.25)^2 + (2\pi r + 0.25)h = 2\pi(r + 0.25)^2 + 2000/r + 250\pi/r^2$$

Thus,

$$M'(r) = 4\pi(r + 0.25) - 2000/r^2 - 500/(\pi r^3).$$

Solving $M'(r) = 0$ for r gives $r = 5.363858$.

12. a) $3.03696, -2.05448, 0.99904486 \pm 0.082161358i$
    b) $-5.07728, 0.84288, -1.7341 \pm 1.0158i, 1.2152 \pm 0.17662i$

14. a) $T_2(x) = 2x^2 - 1$, $T_3(x) = 4x^3 - 3x$, $T_4(x) = 8x^4 - 8x^2 + 1$, $T_5(x) = 16x^5 - 20x^3 + 5x$

    b)   ROOTS

T_3	$0, 0.8660254, -0.8660254$
T_4	$-0.9238795, -0.3826834, 0.3826834, 0.9238795$
T_5	$-0.9510565, -.5877853, 0, 0.5877853, 0.9510565$

16. a) $H_2(x) = 4x^2 - 2$, $H_3(x) = 8x^3 - 12x$, $H_4(x) = 16x^4 - 48x^2 + 12$, $H_5(x) = 32x^5 - 160x^3 + 120x$

    b)   ROOTS

H_3	$-1.224745, 0, 1.224745$
H_4	$-1.650680, -0.5246476, 0.5246476, 1.650680$
H_5	$-2.020183, -0.9585740, 0, 0.9585740, 2.020183$

16

Chapter 3
Interpolation and Polynomial Approximation

Exercise Set 3.1 (PAGE 90)

2. $P_3(x) = 1 - 2x + 3x^2 - 4x^3$, $f(0.05) \simeq 0.907$, $|R_3(0.05)| \leq 3.13 \times 10^{-5}$

4. $n \geq 13$

6. $P_4(x) = x - \frac{1}{2}x^2 + \frac{1}{3}x^3 - \frac{1}{4}x^4$, $\ln(1.1) \simeq 0.0953083$, $|R_4(0.1)| \leq 2 \times 10^{-6}$

8. a) Using the Taylor polynomial of degree three,

$$\ln(x+1) = x - \frac{1}{2}x^2 + \frac{1}{3}x^3 - \frac{x^4}{4(1+\xi)^4} \quad \text{for some } \xi,\ 0 < \xi < x.$$

Since

$$\frac{-x^4}{4(1+\xi)^4} < 0,$$

we have

$$\ln(x+1) < x - \frac{1}{2}x^2 + \frac{1}{3}x^3.$$

Using the Taylor polynomial of degree two,

$$\ln(x+1) = x - \frac{1}{2}x^2 + \frac{x^3}{3(1+\xi)^3} \quad \text{for some } \xi,\ 0 < \xi < x.$$

Since

$$\frac{x^3}{3(1+\xi)^3} > 0,$$

we have

$$\ln(1+x) > x - \frac{1}{2}x^2.$$

b) From part (a)

$$\frac{x}{n} - \frac{x^2}{2n} < \ln\left(1 + \frac{x}{n}\right) < \frac{x}{n} - \frac{x^2}{2n^2} + \frac{x^3}{3n^3}.$$

Thus,

$$-\frac{x^2}{2} < -nx + n^2 \ln\left(1 + \frac{x}{n}\right) < -\frac{x^2}{2} + \frac{x^3}{3n}$$

and

$$e^{-\frac{x^2}{2}} < e^{-nx}e^{n^2 \ln(1+\frac{x}{n})} < e^{-\frac{x^2}{2}}e^{\frac{x^3}{3n}},$$

so

$$e^{-\frac{x^2}{2}} < e^{-nx}\left(1 + \frac{x}{n}\right)^{n^2} < e^{-\frac{x^2}{2}}e^{\frac{x^3}{3n}}.$$

17

Taking the infinity limit on n gives

$$e^{-\frac{x^2}{2}} \leq \lim_{n\to\infty} e^{-nx}\left(1+\frac{x}{n}\right)^{n^2} \leq e^{-\frac{x^2}{2}} \lim_{n\to\infty} e^{\frac{x^3}{3n}} = e^{-\frac{x^2}{2}}$$

10.

$$\frac{k}{n}\binom{n}{k} = \frac{k}{n}\frac{n!}{k!(n-k)!} = \frac{(n-1)!}{(k-1)!(n-k)!} = \frac{(n-1)!}{(k-1)!((n-1)-(k-1))!} = \binom{n-1}{k-1}$$

12.

$$|B_n(x) - x^2| = \left|\frac{n-1}{n}x^2 + \frac{1}{n}x - x^2\right| = \frac{1}{n}|x^2 - x|.$$

Thus,

$$\max_{0\leq x\leq 1} |B_n(x) - x^2| = \frac{1}{4n}.$$

To have $\frac{1}{4n} \leq 10^{-6}$ requires $n \geq 250,000$.

Exercise Set 3.2 (PAGE 98)

2. 0.3334893, error bound 1.3×10^{-6}

4. 0.8094418, error bound 1.30×10^{-4}, actual error -1.18×10^{-4}

6. 0.7137, error bound 2.7×10^{-8}, any discrepancy is due to using only four digits for $\cos x$, and the error formula assumes exact values for $\cos x$.

8. a) 0.2500000

   b) 0.87500000, -4.333333

   c) For a) the bound gives 3.75.

   For b) the bound cannot be used since $f(2)$ is undefined.

10. 0.004

12.

$$w'(x) = \sum_{\substack{i=0}}^{n}\prod_{\substack{j=0\\j\neq i}}^{n}(x - x_j), \quad w'(x_k) = \prod_{\substack{j=0\\j\neq k}}^{n}(x_k - x_j)$$

$$L_k(x) = \frac{\prod_{\substack{j=0 \\ j \neq k}}^{n}(x - x_j)}{\prod_{\substack{j=0 \\ j \neq k}}^{n}(x_k - x_j)} = \frac{\frac{w(x)}{(x-x_k)}}{w'(x_k)} = \frac{w(x)}{(x - x_k)w'(x_k)}$$

14. Since $g'(x - (j + \frac{1}{2})h) = 0$,

$$\max |g(x)| = \max \{|g(x - jh)|, |g(x - (j + \frac{1}{2}))|, |g(x - (j + 1)h)|\}.$$

So

$$|g(x)| = |g(x - (j + \frac{1}{2})h)| = \frac{h^2}{4}.$$

Exercise Set 3.3 (PAGE 104)

2. $1.708\overline{3}$

4. 0.2826

6. a) 0.8094418   b)0.8092831

8. $Q_{i,j} = [(x - x_{j-1})Q_{i,j-1} - (x - x_i)Q_{j-1,j-1}]/(x_i - x_{j-1})$

10. 0.2826

12. 0.038462, 0.333671, 0.116605, −0.371760, −0.0548919, 0.605935, 0.190249, −0.513353, −0.0668173, 0.448335,
    Since $f(1 + \sqrt{10}) = 0.0545716$, the sequence does not appear to converge.

14. Actual Value − Computed Value = $\frac{2}{3}$.

16. Change Algorithm 3.1 as follows:
    INPUT numbers $y_0, y_1, ..., y_n$; values $x_0, x_1, ..., x_n$ as the first column $Q_{0,0}, Q_{1,0}, ..., Q_{n,0}$
        of Q.
    OUTPUT the table Q with $f^{-1}(0)$ approximating $Q_{n,n}$.
    STEP 1 For $i = 1, 2, ..., n$
                for $j = 1, 2, ..., i$
                set
    $$Q_{i,j} = \frac{y_i Q_{i-1,j-1} - y_{i-j}Q_{i,j-1}}{y_i - y_{i-j}}.$$

Exercise Set 3.4 (Page 114)

2. Add $0.01415945x(x - 0.1)(x - 0.3)(x - 0.6)(x - 1)$ to the answer in Exercise 1.

4. 1.1955505

6. 1140

8. $-\frac{11}{12}$

10. a) 192,407,000   b) 571,329,000

11. a)

$$\delta^2 f(x) = \delta f\left(x + \frac{h}{2}\right) - \delta f\left(x - \frac{h}{2}\right)$$
$$= f(x + h) - f(x) - f(x) + f(x - h) = f(x + h) - 2f(x) + f(x - h)$$

b)

$$\delta f\left(x_k + \frac{h}{2}\right) = f\left(x_k + \frac{h}{2} + \frac{h}{2}\right) - f\left(x_k + \frac{h}{2} - \frac{h}{2}\right)$$
$$= f(x_{k+1}) - f(x_k) = hf[x_k, x_{k+1}]$$

and

$$\delta f\left(x_k - \frac{h}{2}\right) = f\left(x_k - \frac{h}{2} + \frac{h}{2}\right) - f\left(x_k - \frac{h}{2} - \frac{h}{2}\right)$$
$$= f(x_k) - f(x_k - h) = f(x_k) - f(x_{k-1}) = hf[x_{k-1}, x_k]$$

Also

$$\delta^2 f(x_k) = f(x_k + h) - 2f(x_k) + f(x_k - h)$$
$$= f(x_{k+1}) - 2f(x_k) + f(x_{k-1}) = 2!h^2 f[x_{k-1}, x_k, x_{k+1}]$$

c) We will first prove that the formula holds for even positive integers and then apply this result to the case when the integer is odd. Note that by part b),

$$\delta^{2m} f(x_k) = h^{2m} (2m)! f[x_{k-m}, \ldots, x_{k+m}]$$

is true for $m = 1$.

Now assume that the formula holds for an arbitrary positive integer m. Then for the integer $m + 1$ we have

$$\delta^{2m+2}f(x_k) = \delta^{2m+1}f\left(x_k + \frac{h}{2}\right) - \delta^{2m+1}f\left(x_k - \frac{h}{2}\right)$$

$$= \delta^{2m}f\left(x_k + \frac{h}{2} + \frac{h}{2}\right) - \delta^{2m}f\left(x_k + \frac{h}{2} - \frac{h}{2}\right)$$

$$- \delta^{2m}f\left(x_k - \frac{h}{2} + \frac{h}{2}\right) + \delta^{2m}f\left(x_k - \frac{h}{2} - \frac{h}{2}\right)$$

$$= \delta^{2m}f(x_{k+1}) - 2\delta^{2m}f(x_k) + \delta^{2m}f(x_{k-1}).$$

Applying the induction hypothesis gives

$$\delta^{2m+2}f(x_k) = h^{2m}(2m!)\Big(f[x_{k+1-m}, \ldots, x_{k+1+m}] - 2f[x_{k-m}, \ldots, x_{k+m}]$$

$$+ f[x_{k-1-m}, \ldots, x_{k-1+m}]\Big)$$

$$= h^{2m}(2m!)\{f[x_{k+1-m}, \ldots, x_{k+1+m}] - f[x_{k-m}, \ldots, x_{k+m}]$$

$$- f[x_{k-m}, \ldots, x_{k+m}] + f[x_{k-1-m}, \ldots, x_{k-1+m}]\}$$

$$= h^{2m}(2m!)\{(x_{k+m+1} - x_{k-m})f[x_{k-m}, \ldots, x_{k+1+m}]$$

$$- (x_{k+m} - x_{k-m-1})f[x_{k-m-1}, \ldots, x_{k+m}]\}$$

$$= h^{2m}(2m!)\{(2m+1)hf[x_{k-m}, \ldots, x_{k+1+m}] - (2m+1)hf[x_{k-m-1}, \ldots, x_{k+m}]\}$$

$$= h^{2m+1}(2m+1)!\big[f[x_{k-m}, \ldots, x_{k+1+m}] - f[x_{k-m-1}, \ldots, x_{k+m}]\big]$$

$$= h^{2m+1}(2m+1)!\big[(x_{k+1+m} - x_{k-m-1})f[x_{k-m-1}, \ldots, x_{k+m+1}]\big]$$

$$= h^{2m+2}(2m+2)!f[x_{k-m-1}, \ldots, x_{k+m+1}].$$

By the principle of mathematical induction this implies that the formula holds for arbitrary positive even integers, that is, integers of the form $2m$.

Since the formula holds for arbitrary positive even integers, for arbitrary positive odd integers we have

$$\delta^{2m+1}f\left(x_k + \frac{h}{2}\right) = \delta^{2m}f\left(x_k + \frac{h}{2} + \frac{h}{2}\right) - \delta^{2m}f\left(x_k + \frac{h}{2} - \frac{h}{2}\right)$$

$$= \delta^{2m}f(x_{k+1}) - \delta^{2m}f(x_k)$$

$$= h^{2m}(2m)!f[x_{k+1-m}, \ldots, x_{k+1+m}] - h^{2m}(2m)!f[x_{k-m}, \ldots, x_{k+m}]$$

$$= h^{2m}(2m)!\Big((x_{k+1+m} - x_{k-m})f[x_{k-m}, \ldots, x_{k+1+m}]\Big)$$

$$= h^{2m+1}(2m+1)!f[x_{k-m}, \ldots, x_{k+1+m}]$$

and

$$\delta^{2m+1} f\left(x_k - \frac{h}{2}\right) = \delta^{2m} f(x_k) - \delta^{2m} f(x_{k-1})$$

$$= h^{2m}(2m)! \left[f[x_{k-m}, \ldots , x_{k+m}] - f[x_{k-1-m}, \ldots , x_{k-1+m}] \right]$$

$$= h^{2m}(2m)! \left[(x_{k+m} - x_{k-1-m}) f[x_{k-1-m}, \ldots , x_{k+m}] \right]$$

$$= h^{2m+1}(2m+1)! \, f[x_{k-1-m}, \ldots , x_{k+m}].$$

Hence the formula holds for arbitrary positive odd integers as well.

12. a) First note that

$$\mu\delta f(x_k) = \frac{1}{2}\left[\delta f\left(x_k + \frac{h}{2}\right) + \delta f\left(x_k - \frac{h}{2}\right) \right]$$

$$= \frac{1}{2}[f(x_{k+1}) - f(x_k) + f(x_k) - f(x_{k-1})]$$

$$= \frac{1}{2}[f(x_{k+1}) - f(x_{k-1})]$$

and that

$$\frac{1}{2}\left[\delta f\left(x_k + \frac{h}{2}\right) + \delta f\left(x_k - \frac{h}{2}\right) \right] = \frac{f(x_{k+1}) - f(x_k)}{2} + \frac{f(x_k) - f(x_{k-1})}{2}$$

$$= \frac{1}{2}[f(x_{k+1}) - f(x_{k-1})],$$

so

$$\mu\delta f(x_k) = \frac{1}{2}\left[\delta f\left(x_k + \frac{h}{2}\right) + \delta f\left(x_k - \frac{h}{2}\right) \right].$$

b) Since

$$\mu\delta^3 f(x_k) = \mu\left[\delta^2 f\left(x_k + \frac{h}{2}\right) - \delta^2 f\left(x_k - \frac{h}{2}\right) \right]$$

$$= \mu\left[f\left(x_{k+1} + \frac{h}{2}\right) - 2f\left(x_k + \frac{h}{2}\right) + f\left(x_k - \frac{h}{2}\right) - f\left(x_k + \frac{h}{2}\right)\right.$$

$$\left. + 2f\left(x_k - \frac{h}{2}\right) - f\left(x_{k-1} - \frac{h}{2}\right) \right]$$

$$= \mu\left[f\left(x_{k+1} + \frac{h}{2}\right) - 3f\left(x_k + \frac{h}{2}\right) + 3f\left(x_k - \frac{h}{2}\right) - f\left(x_{k-1} - \frac{h}{2}\right) \right]$$

$$= \frac{1}{2}\left[f(x_{k+2}) + f(x_{k+1}) \right] - \frac{3}{2}\left[f(x_{k+1}) + f(x_k) \right] + \frac{3}{2}\left[f(x_k) + f(x_{k-1}) \right]$$

$$- \frac{1}{2}\left[f(x_{k-1}) + f(x_{k-2}) \right]$$

$$= \frac{1}{2}\left[f(x_{k+2}) - 2f(x_{k+1}) + 2f(x_{k-1}) - f(x_{k-2}) \right]$$

and

$$\frac{1}{2}\left[\delta^3 f\left(x_k + \frac{h}{2}\right) + \delta^3 f\left(x_k - \frac{h}{2}\right)\right] = \frac{1}{2}\left[f(x_k + 2h) - 3f(x_k + h) + 3f(x_k) - f(x_k - h)\right]$$

$$+ \frac{1}{2}\left[f(x_k + h) - 3f(x_k) + 3f(x_{k-1}) - f(x_k - 2h)\right]$$

$$= \frac{1}{2}\left[f(x_{k+2}) - 2f(x_{k+1}) + 2f(x_{k-1}) - f(x_{k-2})\right],$$

we have

$$\frac{1}{2}\left[\delta^3 f\left(x_k + \frac{h}{2}\right) + \delta^3 f\left(x_k - \frac{h}{2}\right)\right] = \frac{1}{2}\left[f(x_{k+2}) - 2f(x_{k+1}) + 2f(x_{k-1}) - f(x_{k-2})\right].$$

13. Stirling's formula is given by

$$P_n(x) = P_{2n+1}(x) = f[x_0] + \frac{sh}{2}(f[x_{-1}, x_0] + f[x_0, x_1]) + s^2 h^2 f[x_{-1}, x_0, x_1]$$

$$+ \frac{s(s^2 - 1)h^3}{2}(f[x_{-1}, x_0, x_1, x_2] + f[x_{-2}, x_{-1}, x_0, x_1])$$

$$+ \cdots + s^2(s^2 - 1)(s^2 - 4)\ldots(s^2 - (m-1)^2)h^{2m} f[x_{-m}, \ldots, x_m]$$

$$+ \left\{\frac{s(s^2 - 1)\ldots(s^2 - m^2)h^{2m+1}}{2}(f[x_{-m}, \ldots, x_{m+1}] + f[x_{-m-1} \ldots, x_m])\right\}.$$

But,

$$h(f[x_{-1}, x_0] + f[x_0, x_1]) = hf[x_{-1}, x_0] + hf[x_0, x_1]$$

$$= \delta f(x_0 - \frac{h}{2}) + \delta f(x_0 + \frac{h}{2}).$$

Also,

$$s^2 h^2 f[x_{-1}, x_0, x_1] = s^2 h^2 \left[\frac{\delta^2 f(x_0)}{h^2 2}\right] = \frac{s^2}{2}\delta^2 f(x_0),$$

$$\frac{s(s^2 - 1)}{2}h^3\left(f[x_{-1}, x_0, x_1, x_2] + f[x_{-2}, x_{-1}, x_0, x_1]\right)$$

$$= \frac{s(s^2 - 1)}{2}h^3\left[\frac{\delta^2 f(x_0 + \frac{h}{2})}{h^3 3!} + \frac{\delta^3 f(x_0 - \frac{h}{2})}{h^3 3!}\right]$$

$$= \frac{s(s^2 - 1)}{2 \cdot 3!}\left[\delta^3 f(x_0 + \frac{h}{2}) + \delta^3 f(x_0 - \frac{h}{2})\right],$$

$$s^2(s^2 - 1)h^2 f[x_{-2}, \ldots, x_2] = s^2(s^2 - 1)h^2\left[\frac{\delta^4 f(x_0)}{h^2 4!}\right] = \frac{s^2(s^2 - 1)}{4!}\delta^4 f(x_0),$$

23

and

$$s^2(s^2 - 1^2)(s^2 - 2^2)\cdots(s^2 - (m-1)^2)h^{2m}f[x_{-m}, \ldots, x_m]$$

$$= s^2(s^2 - 1^2)(s^2 - 2^2)\cdots(s^2 - (m-1)^2)h^{2m}\left[\frac{\delta^{2m}f(x_0)}{h^{2m}(2m)!}\right]$$

$$= \frac{s^2(s^2 - 1)(s^2 - 2^2)\cdots[s^2 - (m-1)^2)]}{(2m)!}\delta^{2m}f(x_0).$$

Finally,

$$\frac{s(s^2 - 1)\cdots(s^2 - m^2)h^{2m+1}}{2})(f[x_{-m}, \ldots, x_{m+1}] + f[x_{-m-1}, \ldots, x_m])$$

$$= \frac{s(s^2 - 1)\cdots(s^2 - m^2)}{2(2m+1)!}\left(\delta^{2m+1}f\left(x_0 + \frac{h}{2}\right) + \delta^{2m+1}f\left(x_0 - \frac{h}{2}\right)\right)$$

This demonstrates, term by term, the equivalence of the two formulas.

14. Compare the terms to those in Exercise 13.

16. From Eq. (3.5),

$$f(x) = P_n(x) + \frac{f^{n+1}(\xi(x))}{(n+1)!}(x - x_0)\ldots(x - x_n).$$

Let $x_{n+1} = x$. The interpolation polynomial of degree $n + 1$ on $x_0, ..., x_{n+1}$ is given by

$$P_{n+1}(t) = P_n(t) + f[x_0, x_1, ..., x_n, x_{n+1}](t - x_0)(t - x_1)\ldots(t - x_n).$$

Since $f(x) = f(x_{n+1}) = P_{n+1}(x)$,

$$P_n(x) + \frac{f^{n+1}(\xi(x))}{(n+1)!}(x - x_0)\ldots(x - x_n) = P_n(x) + f[x_0, ..., x_n, x](x - x_0)\ldots(x - x_n).$$

Thus,

$$f[x_0, ..., x_n, x] = \frac{f^{n+1}(\xi(x))}{(n+1)!}.$$

24

Exercise Set 3.5 (PAGE 124)

2. $H(1.03) = 0.80932485$, | actual error $| = 1.24 \times 10^{-6}$, | error bound $| \leq 1.31 \times 10^{-6}$.

4. $H(1.03)$ is better, $|P_2(1.03) - f(1.03)| \leq 1.30 \times 10^{-4}$, $|H_3(1.03) - f(1.03)| \leq 1.31 \times 10^{-6}$.

6. a) 0.33349

b) 3.05×10^{-14}. The actual error is 2.91×10^{-6} due to rounding.

8. a) Suppose P is another polynomial with $P(x_k) = f(x_k)$ and $P'(x_k) = f'(x_k)$ for $k = 0, ..., n$ and the degree of P is at most $2n + 1$. Let

$$D(x) = H_{2n+1}(x) - P(x).$$

Then D is a polynomial of degree at most $2n + 1$ with $D(x_k) = 0$ and $D'(x_k) = 0$ for each $k = 0, 1, ..., n$. Thus, $D(x)$ has zeros of multiplicity 2 at each x_k and

$$D(x) = (x - x_0)^2 \ldots (x - x_n)^2 Q(x).$$

Hence, D must be of degree $2n$ or more, which would be a contradiction, or $Q(x) \equiv 0$ which implies that $D(x) \equiv 0$. Thus, $P(x) \equiv H_{2n+1}(x)$.

b) The error formula holds if $x = x_k$ for any choice of ξ.

Let $x \neq x_k$ for $k = 0, ..., n$ and define

$$g(t) = f(t) - H_{2n+1}(t) - \frac{(t - x_0)^2 \ldots (t - x_n)^2}{(x - x_0)^2 \ldots (x - x_n)^2} [f(x) - H_{2n+1}(x)].$$

Note that $g(x_k) = 0$ for $k = 0, ..., n$ and $g(x) = 0$. Thus, g has $n + 2$ distinct zeros in $[a, b]$. By Rolle's Theorem, g' has $n+1$ distinct zeros, $\xi_0, ..., \xi_n$ which are between the numbers $x_0, ..., x_n, x$.

In addition, $g'(x_k) = 0$ for $k = 0, ..., n$, so g' has $2n + 2$ distinct zeros $\xi_0, ..., \xi_n, x_0, ..., x_n$. Since g' is $2n + 1$ times differentiable, the Generalized Rolle's Theorem implies that a number ξ in $[a, b]$ exists with $g^{(2n+2)}(\xi) = 0$.

But,

$$g^{(2n+2)}(t) = f^{(2n+2)}(t) - \frac{d^{2n+2}}{dt^{2n+2}} H_{2n+1}(t) - [f(x) - H_{2n+1}(x)] \cdot (2n + 2)!$$

and

$$0 = g^{(2n+2)}(\xi) = f^{(2n+2)}(\xi) - (2n + 2)![f(x) - H_{2n+1}(x)].$$

The error formula follows immediately.

10. The Hermite polynomial predicts a distance of 742 feet and a speed of 48 feet per second. The speed prediction demonstrates the undesirability of using polynomials for derivative approximation.

25

Exercise Set 3.6 (Page 138)

2. a) 0.4980696   **b)** 1.621986   **c)** -3.7163×10^{-5} **d)** 1.16528   **e)** 0.8590474
f) 1.531698

4. The equation of the spline is

$$S(x) = S_i(x) = a_i + b_i(x - x_i) + c_i(x - x_i)^2 + d_i(x - x_i)^3$$

on the interval $[x_i, x_{i+1}]$ where the results are given in the following table:

x_i	a_i	b_i	c_i	d_i
0	1.00000	−0.923601	0	0.620865
0.25	0.778801	−0.807189	0.465649	−0.154016
0.75	0.472367	−0.457052	0.234624	−0.312832

$\int_0^1 S(x)\,dx = 0.631967$, $S'(0.5) = -0.603243$, $S''(0.5) = 0.700274$.

6. The equation of the spline is

$$S(x) = S_i(x) = a_i + b_i(x - x_i) + c_i(x - x_i)^2 + d_i(x - x_i)^3$$

on the interval $[x_i, x_{i+1}]$ where the results are given in the following table:

x_i	a_i	b_i	c_i	d_i
0	1.00000	−1.00000	0.499440	−0.154515
0.25	0.778801	−0.779251	0.383555	−0.101580
0.75	0.472367	−0.471881	0.231185	−0.0618174

$$\int_0^1 S(x)\,dx = 0.623078, \quad S'(0.5) = -0.606520, \quad \text{and} \quad S''(0.5) = 0.614740.$$

8. a) 0.33348   **b)** To 5 digits, $\sin 0.34 = 0.33349$.
c) 0.33349   **d)** To 5 digits, $\sin 0.34 = 0.33349$.
e) 0.94274   **f)** 0.94324
g) 0.015964   **h)** 0.015964

10. a) On $[0, 0.05]$ $s(x) = 1.000000 + 1.999999x + 1.998302x^2 + 1.401310x^3$.

On $[0.05, 0.1]$

$$s(x) = 1.105170 + 2.210340(x - 0.05) + 2.208498(x - 0.05)^2 + 1.548758(x - 0.05)^3.$$

12. The equation of the spline is

$$S(x) = S_i(x) = f_i + b_i(x - x_i) + c_i(x - x_i)^2 + d_i(x - x_i)^3$$

on the interval $[x_i, x_{i+1}]$ where the results are given in the following table.

x_i	f_i	b_i	c_i	d_i
0.9	0.7	3.541214	−23.11823	77.06073
0.8	0.3	2.605763	32.47269	−185.3031
0.7	0.2	1.035753	−16.77264	164.1511
0.6	0.0	2.251228	4.617891	−71.3018
0.5	−0.2	1.959329	−1.698893	21.05594
0.4	−0.4	1.911449	2.177681	−12.92191
0.3	−0.5	−0.6051273	22.98808	−69.36810
0.2	−0.3	−2.490925	−4.130146	90.39406
0.1	−0.1	−1.431164	−6.467466	7.79107
0	0	−0.7844177	0	−21.55821

13. Before STEP 7 in Algorithm 3.4 and STEP 8 in Algorithm 3.5 insert the following:

For $j = 0, \ldots, n - 1$ set

$\quad\quad l_1 = b_j$; (*Note that* $l_1 = s'(x_j)$.)

$\quad\quad l_2 = 2c_j$; (*Note that* $l_2 = s''(x_j)$.)

$\quad\quad$ OUTPUT (l_1, l_2)

Set

$\quad\quad l_1 = b_{n-1} + 2c_{n-1}h_{n-1} + 3d_{n-1}h_{n-1}^2$; (*Note that* $l_1 = s'(x_n)$.)

$\quad\quad l_2 = 2c_{n-1} + 6d_{n-1}h_{n-1}$; (*Note that* $l_2 = s''(x_n)$.)

$\quad\quad$ OUTPUT (l_1, l_2).

14. Before STEP 7 in Algorithm 3.2 and STEP 5 in Algorithm 3.3 insert the following:

Set $I = 0$

For $j = 0, \ldots, n - 1$ set

$\quad\quad I = a_j h_j + \frac{b_j}{2} h_j^2 + \frac{c_j}{3} h_j^3 + \frac{d_j}{4} h_j^4 + I.$ (*Accumulate* $\int_{x_j}^{x_{j+1}} S(x)\, dx.$)

OUTPUT (I).

16. The five equations are

$$a_0 = f(x_0), \quad a_1 = f(x_1), \quad a_1 + b_1(x_2 - x_1) + c_1(x_2 - x_1)^2 = f(x_2),$$
$$a_0 + b_0(x_1 - x_0) + c_0(x_1 - x_0)^2 = a_1, \quad \text{and} \quad b_0 + 2c_0(x_1 - x_0) = b_1.$$

If $S \in C^2$, then S is a quadratic on $[x_0, x_2]$ and the solution may not be meaningful.

18. To show that $s'' \in C[a, b]$ it suffices to show that s'' is continuous at x_j for $j = 1, ..., n-1$. We first define simplifying notation for the right- and left-sided limits of s at a:

$$s(a+) = \lim_{x \to a^+} s(x) \quad \text{and} \quad s(a-) = \lim_{x \to a^-} s(x).$$

Using this notation we have

$$s''(x_j-) = s''_{j-1}(x_j) = a_j = s''_j(x_j) = s''(x_j+),$$

so s'' is continuous at x_j. The first integration gives

$$s'_j(x) = \frac{a_{j+1}(x - x_j)^2}{2h_j} - \frac{a_j(x_{j+1} - x)^2}{2h_j} + c$$

and the second results in

$$s_j(x) = \frac{a_{j+1}(x - x_j)^3}{6h_j} + \frac{a_j(x_{j+1} - x)^3}{6h_j} + cx + c_1.$$

To obtain c and c_1, note that

$$f_j = s_j(x_j) = \frac{a_j h_j^2}{6} + cx_j + c_1$$

and

$$f_{j+1} = s_j(x_{j+1}) = \frac{a_{j+1} h_j^2}{6} + cx_{j+1} + c_1.$$

Solving for c and c_1 gives the required formula. Since s' is continuous at x_j it follows that

$$s'_j(x_j) = s'_{j-1}(x_j)$$

so

$$\frac{-a_j h_j}{2} - \frac{a_{j+1} h_j}{6} + \frac{a_j h_j}{6} + \frac{f_{j+1} - f_j}{h_j} = \frac{a_j h_{j-1}}{2} - \frac{a_j h_{j-1}}{6} + \frac{a_{j-1} h_{j-1}}{6} + \frac{f_j - f_{j-1}}{h_{j-1}}.$$

Collecting terms and simplifying gives, for each $j = 1, \ldots, n-1$,

$$h_{j-1} a_{j-1} + 2(h_{j-1} + h_j)a_j + h_j a_{j+1} = \frac{6}{h_j}(f_{j+1} - f_j) - \frac{6}{h_{j-1}}(f_j - f_{j-1}).$$

Two additional conditions,

either

$$s''(a) = s''(b) = 0$$

or

$$s'(a) = f'(a) \quad \text{and} \quad s'(b) = f'(b),$$

supply two more equations. Thus, $n+1$ equations in the $n+1$ unknowns $a_0, a_1, ..., a_n$ must be solved to obtain the cubic spline.

19. On $[x_0, x_1]$, $\quad s(x) = 0$, $s'(x) = 0$ and $s''(x) = 0$.

On $[x_1, x_2]$, $s(x) = c_1(x - x_1)^3$, $s'(x) = 3c_1(x - x_1)^2$ and $s''(x) = 6c_1(x - x_1)$.

On $[x_2, x_3]$, $s(x) = c_1(x - x_1)^3 + c_2(x - x_2)^3$, $s'(x) = 3c_1(x - x_1)^2 + 3c_2(x - x_2)^2$ and $s''(x) = 6c_1(x - x_1) + 6c_2(x - x_2)$.

Using the notation for right- and left-sided limits introduced in the solution to Exercise 18, we have

$$s(x_1-) = s(x_1+) = 0 \quad \text{and} \quad s(x_2-) = s(x_2+) = c_1(x_2 - x_1)^3.$$

Similarly,

$$s'(x_1-) = s'(x_1+) = 0, \quad s'(x_2-) = s'(x_2+) = 3c_1(x_2 - x_1)^2$$

$$s''(x_1-) = s''(x_1+) = 0, \quad \text{and} \quad s''(x_2-) = s''(x_2+) = 6c_1(x_2 - x_1).$$

Hence, $s \in C^2[x_0, x_3]$.

20. On $[x_{i-1}, x_i]$ we have

$$S(x) = S_{i-1}(x) = P(x) + \sum_{j=1}^{i-1} c_j(x - x_j)^3$$

$$S'(x) = S'_{i-1}(x) = P'(x) + 3\sum_{j=1}^{i-1} c_j(x - x_j)^2$$

$$S''(x) = S''_{i-1}(x) = P''(x) + 6\sum_{j=1}^{i-1} c_j(x - x_j)$$

and on $[x_i, x_{i+1}]$

$$S(x) = S_i(x) = P(x) + \sum_{j=1}^{i} c_j(x - x_j)^3$$

$$S'(x) = S_i'(x) = P'(x) + 3\sum_{j=1}^{i} c_j(x - x_j)^2$$

$$S''(x) = S_i''(x) = P''(x) + 6\sum_{j=1}^{i} c_j(x - x_j).$$

Using these equations and the notation for right- and left-sided limits introduced in the solution to Exercise 18, we have

$$S(x_i-) = S_{i-1}(x_i) = P(x_i) + \sum_{j=1}^{i-1} c_j(x_i - x_j)^3 = S_i(x_i) = S(x_i+),$$

$$S'(x_i-) = S_{i-1}'(x_i) = P'(x_i) + 3\sum_{j=1}^{i-1} c_j(x_i - x_j)^2 = S_i'(x_i) = S'(x_i+),$$

$$S''(x_i-) = S_{i-1}''(x_i) = P''(x_i) + 6\sum_{j=1}^{i-1} c_j(x_i - x_j) = S_i''(x_i) = S''(x_i+).$$

Thus, for each $i = 1, ..., n - 1, S, S', S''$ are continuous at x_i. Hence, $S \in C^2[a, b]$ since S is a cubic polynomial on $[x_i, x_{i+1}]$ for each $i = 0, 1, ..., n - 1$. This implies that S is a cubic spline.

21. Consider the intervals $[x_{i-1}, x_i]$ and $[x_i, x_{i+1}]$.

On $[x_{i-1}, x_i]$,

$$s(x) = P(x) + \sum_{j=1}^{i-1} c_j(x - x_j)^3$$

so

$$s'''(x_i-) = P'''(x_i) + \sum_{j=1}^{i-1} 6c_j.$$

and on $[x_i, x_{i+1}]$,

$$s(x) = P(x) + \sum_{j=1}^{i} c_j(x - x_j)^3$$

so

$$s'''(x_i+) = P'''(x_i) + \sum_{j=1}^{i} 6c_j.$$

If $s'''(x_i-) = s'''(x_i+)$, then $c_i = 0$.

22. $S(x) = S_i(x) = a_i + b_i(x - x_i) + c_i(x - x_i)^2 + d_i(x - x_i)^3$
on $[x_i, x_{i+1}]$ where

x_i	a_i	b_i	c_i	d_i
0	0	75	-0.659292	0.219764
3	225	76.99779	1.31858	-0.153761
5	383	80.4071	0.396018	-0.177237
8	623	77.9978	-1.19912	0.0799115

$S(10) = 774.84$, $S'(10) = 74.16$ ft/sec
Max $S'(x) = S'(5.7448) = 80.7$ ft/sec = 55.02 MPH.
55 MPH was first exceeded at 5.5 sec.

24. a) $S(x) = S_i(x) = a_i + b_i(x - x_i) + c_i(x - x_i)^2 + d_i(x - x_i)^3$ on $[x_i, x_{i+1}]$ where

x_i	a_i	b_i	c_i	d_i
0	0	103.0425	0	-23.0820
0.25	25.4	98.7147	-17.3114	25.8098
0.5	49.4	94.8984	2.04590	1.91475
1.0	97.6	98.3803	4.91803	-6.55738
1.25				

b) $1{:}13\frac{7}{25}$
c) Starting speed $\simeq 9.7047 \times 10^{-3}$ miles/sec = 34.94 MPH.
   Ending speed $\simeq 36.14$ MPH.

Chapter 4
Numerical Differentiation and Integration

Exercise Set 4.1 (PAGE 154)

2. a) -0.1951027 b) -1.541415 c) -0.6824175

4.

FORMULA	$h = 0.1$	$h = 0.01$
(4.12)	5782.906	7300.911
(4.13)	7774.581	7314.993

6. -0.60645, error bound 0.00013.

8. 2.270, 2.300

10. $h = 0.1$, 36.641; $h = 0.01$, 36.5. The actual value is 36.5935.

12. -0.4249840, -1.032772

14. $f'(x_0) = \frac{1}{6h}[-8f(x_0) + 9f(x_0 + h) - f(x_0 + 3h)]$, 4495.540, 7295.904

16. Optimal $h = 2\sqrt{\frac{\varepsilon}{M}}$, where $M = \max |f''(x)|$.

18. $e'(h) = -\frac{\varepsilon}{h^2} + \frac{hM}{3}$, $e'(h) = 0$ if $h = \sqrt[3]{3\varepsilon/M}$.
  Since $e'(h) < 0$ if $h < \sqrt[3]{3\varepsilon/M}$ and $e'(h) > 0$ if $h > \sqrt[3]{3\varepsilon/M}$, an absolute minimum for $e(h)$ occurs at $h = \sqrt[3]{3\varepsilon/M}$.

20. a) On $[1.02, 1.04]$

$$s'(x) = 1.424342 - 5.634594(x - 1.02) - 14.44860(x - 1.02)^2$$

and $s'(1.03) = 1.366551$.
b) On $[1.02, 1.04]$

$$s'(x) = 1.423630 - 5.40870(x - 1.02) - 26.02125(x - 1.02)^2$$

and $s'(1.03) = 1.366941$.

21. $f'''(x_0) = \frac{1}{h^3}[-\frac{1}{2}f(x_0 - 2h) + f(x_0 - h) - f(x_0 + h) + \frac{1}{2}f(x_0 + 2h)] - \frac{h^2}{240}f^5(\xi)$

22. -0.616067 The actual value is -0.60653. The error is 0.0095.

Exercise Set 4.2 (Page 162)

2. a) 1.0000000   b) 2.0000000   c) 2.2751458   d) −19.646796

4. a) 1.000   b) 1.998   c) 2.281   d)−19.68

6. −1.000135

7. With $h = 0.1$ (4.14) becomes

$$f'(2) \approx \frac{1}{1.2}[1.8e^{1.8} - 8(1.9e^{1.9}) + 8(2.1)e^{2.1} - 2.2e^{2.2}] = 22.166995.$$

With $h = 0.05$, (4.14) becomes

$$f'(2) \approx \frac{1}{0.6}[1.9e^{1.9} - 8(1.95e^{1.95}) + 8(2.05)e^{2.05} - 2.1e^{2.1}] = 22.167157.$$

8. $\frac{1}{12h}[f(x_0 + 4h) - 12f(x_0 + 2h) + 32f(x_0 + h) - 21f(x_0)]$

9. a)
$$P_{0,1}(h) = \frac{(x - h^2)N_1(\frac{h}{2})}{\frac{h^2}{4} - h^2} + \frac{(x - \frac{h^2}{4})N_1(h)}{h^2 - \frac{h^2}{4}}, \text{ so } P_{0,1}(0) = \frac{4N_1(\frac{h}{2}) - N_1(h)}{3}$$

Similarly,
$$P_{1,2}(0) = \frac{4N_1(\frac{h}{4}) - N_1(\frac{h}{2})}{3}.$$

b)
$$P_{0,2}(h) = \frac{(x - h^4)N_2(\frac{h}{2})}{\frac{h^4}{16} - h^4} + \frac{(x - \frac{h^4}{16})N_2(h)}{h^4 - \frac{h^4}{16}},$$

so
$$P_{0,2}(0) = \frac{16N_2(\frac{h}{2}) - N_2(h)}{15}.$$

Exercise Set 4.3 (Page 171)

2. a) 1.4972   b) 1.4775

4. $n = 3$: 0.7669158, error bound 1.872×10^{-4}
   $n = 2$: 0.7665748, error bound 3.686×10^{-4}

6.

(4.40)	(4.41)	(4.43)	(4.46)	(4.48)
5.43476	5.03420	5.03292	4.83393	5.03180

8.

(4.40)	(4.41)	(4.42)	(4.43)	(4.46)	(4.47)	(4.48)	(4.49)
1.0500	0.92143	0.91875	0.91648	0.85714	0.87500	0.91215	0.91330

9. For

$$f(x) = x : a_0 x_0 + a_1(x_0 + h) + a_2(x_0 + 2h) = 2x_0 h + 2h^2;$$

$$f(x) = x^2 : a_0 x_0^2 + a_1(x_0 + h)^2 + a_2(x_0 + 2h)^2 = 2x_0^2 h + 4x_0 h^2 + \frac{8h^3}{3};$$

$$f(x) = x^3 : a_0 x_0^3 + a_1(x_0 + h)^3 + a_2(x_0 + 2h)^3 = 2x_0^3 h + 6x_0^2 h^2 + 8x_0 h^3 + 4h^4.$$

Solving this linear system for a_0, a_1, a_2 gives $a_0 = \frac{h}{3}, a_1 = \frac{4h}{3}$, and $a_2 = \frac{h}{3}$.
Using $f(x) = x^4$ gives $f^{(4)}(\xi) = 24$, so

$$\frac{1}{5}(x_2^5 - x_0^5) = \frac{h}{3}(x_0^4 + 4x_1^4 + x_2^4) + 24k.$$

Replacing x, with $x_0 + h, x^2$ with $x_0 + 2h$, and simplifying gives $k = -h^5/90$.

10. If a quadrature formula has degree of precision n, then $E(x^k) = 0$ for all $k = 0, 1, \ldots, n$ follows immediately since x^k is a polynomial of degree $\leq n$.
Let $P_{n+1}(x) = a_0 x^{n+1} + a_1 x^n + \cdots + a_{n+1}$ be a polynomial of degree $n + 1$ for which $E(P_{n+1}) \neq 0$. Since

$$x^{n+1} = \frac{1}{a_0} P_{n+1}(x) - \frac{a_1}{a_0} x^n - \cdots - \frac{a_{n+1}}{a_0},$$

we have

$$E(x^{n+1}) = \frac{1}{a_0} E(P_{n+1}) - E\left(\frac{a_1}{a_0} x^n + \cdots + \frac{a_{n+1}}{a_0}\right) = \frac{1}{a_0} E(P_{n+1}) \neq 0$$

If $E(x^k) = 0$ for all $k = 0, 1, \ldots, n$ and $E(x^{n+1}) \neq 0$, then, with $P_{n+1}(x) = x^{n+1}$, we have $E(P_{n+1}) \neq 0$. Let $P(x) = a_0 x^n + \cdots + a_n$ be any polynomial of degree $\leq n$. Then $E(P) = a_0 E(x^n) + \cdots + a_n E(x^0) = 0$.

11. Using $n = 3$ in Theorem 4.2 gives

$$\int_a^b f(x)dx = \sum_{i=0}^3 a_i f(x_i) + \frac{h^5 f^{(4)}(\xi)}{24} \int_0^3 t(t-1)(t-2)(t-3)dt.$$

Since

$$\int_0^3 t(t-1)(t-2)(t-3)dt = -\frac{9}{10},$$

the error term is

$$-3h^5 f^{(4)}(\xi)/80.$$

Also,

$$a_i = \int_{x_0}^{x_3} \prod_{\substack{j=0 \\ j \neq i}}^{3} \frac{x - x_j}{x_i - x_j} \, dx \quad \text{for } i = 0, 1, 2, 3.$$

Using the change of variables $x = x_0 + th$ gives

$$a_i = h \int_{0}^{3} \prod_{\substack{j=0 \\ j \neq i}}^{3} \frac{t - j}{i - j} \, dt \quad \text{for } i = 0, 1, 2, 3.$$

Evaluating the integrals gives $a_0 = \frac{3h}{8}, a_1 = \frac{9h}{8}, a_2 = \frac{9h}{8}$, and $a_3 = \frac{3h}{8}$.

12. We have $n = 1$, $x_{-1} = a$, $x_0 = a + h$, $x_1 = x_0 + h$, $x_2 = x_0 + 2h$ so $L_0(x) = -(x - x_1)/h$ and $L_1(x) = (x - x_0)/h$. Thus,

$$a_0 = \int_{x_0-h}^{x_0+2h} \frac{x - x_1}{-h} \, dx = \frac{3h}{2} = a_1 = \int_{x_0-h}^{x_0+2h} \frac{x - x_0}{h} \, dx.$$

Using (4.45) with $n = 1$ gives

$$\text{Error term} = \frac{h^3 f''(\xi)}{2} \int_{-1}^{2} t(t - 1) \, dt = \frac{3}{4} h^3 f''(\xi).$$

Exercise Set 4.4 (PAGE 197)

2. a) 1.083333   b) 3.777778   c) 10.08894   d) 0.6532815   e) −6.716470
   f) 0.7047473

4. a) 0.9163064   b) 0.9178617   c) 0.9141301

6. With $m = 12$, 14.02564

8. a) 0.3466739, 0.3465801   b) 5.1883×10^{-4}, 3.2427×10^{-5} c) $h < 0.013$, $n > 62$

10. a) −0.3497582, −0.3473746   b) 0.0101, 0.0025   c) $n \geq 4019$, $h \leq 0.000196$

12. a) −0.3437928, −0.3455552   b) 0.00898, 0.00330   c) $n \geq 5684$, $h < 0.0001382$

14. Let $h = (b-a)/n$ and $x_j = a + jh$ for each $j = 0, 1, ..., n$. Then

$$\int_a^b f(x)dx = \sum_{j=1}^n \int_{x_{j-1}}^{x_j} f(x)dx$$

$$= \sum_{j=1}^n \left[\frac{h}{2}(f(x_{j-1}) + f(x_j)) - \frac{h^3}{12}f''(\xi_j) \right]$$

$$= \frac{h}{2}\left[f(x_0) + 2\sum_{j=1}^{n-1} f(x_j) + f(x_n) \right] - \frac{h^3}{12}\sum_{j=1}^n f''(\xi_j)$$

$$= \frac{h}{2}\left[f(a) + 2\sum_{j=1}^{n-1} f(x_j) + f(b) \right] - \frac{h^3}{12}\sum_{j=1}^n f''(\xi_j),$$

where $x_{j-1} < \xi_j < x_j$,

But,

$$\min_{x \in [a,b]} f''(x) \le f''(\xi_j) \le \max_{x \in [a,b]} f''(x)$$

so

$$\min_{x \in [a,b]} f''(x) \le \frac{1}{n}\sum_{j=1}^n f''(\xi_j) \le \max_{x \in [a,b]} f''(x).$$

By the Intermediate Value Theorem, there exists $\mu \in (a, b)$ with

$$f''(\mu) = \frac{1}{n}\sum_{j=1}^n f''(\xi_j).$$

So

$$\frac{-h^3}{12}\sum_{j=1}^n f''(\xi_j) = \frac{-h^3 n}{12}f''(\mu) = -\frac{h^2(b-a)}{12}f''(\mu).$$

To implement the Composite Trapezoidal Method, change Algorithm 4.1 as follows:
STEP 1 Set $h = (b-a)/n$.
STEP 2 Set XI0 $= f(a) + f(b)$;
         XI1 $= 0$.
STEP 3 For $i = 1, ..., n-1$ do STEPS 4 and 5.
     STEP 4 Set $x = a + ih$.
     STEP 5 Set XI1 $=$ XI1 $+ f(x)$.
STEP 6 Set XI $= h$(XI0 $+$ 2XI1)/2.

15. Set $h = (b-a)/(2m+2)$ and $x_j = a + (j+1)h$ for each $j = -1, 0, ..., 2m+1$. Then,

$$\int_a^b f(x)dx = \int_{x_{-1}}^{x_{2m+1}} f(x)dx$$

$$= \sum_{j=0}^m \int_{x_{2j-1}}^{x_{2j+1}} f(x)dx$$

$$= 2h\sum_{j=0}^m f(x_{2j}) + \frac{h^3}{3}\sum_{j=0}^m f''(\xi_j),$$

36

where $x_{2j-1} < \xi_j < x_{2j+1}$.

As in Exercise 14, there exists $\mu \in (a, b)$ with

$$(m+1)f''(\mu) = \sum_{j=0}^{m} f(\xi_j).$$

Thus, the error term becomes

$$\frac{h^3(m+1)}{3} = \frac{h^2(b-a)(m+1)}{3(2m+2)} = \frac{h^2(b-a)}{6}.$$

To implement the Composite Midpoint Method, change Algorithm 4.1 as follows:

STEP 1 Set $h = (b-a)/(2m+2)$.

STEP 2 Set XI = 0.

STEP 3 For $i = 0, 1, ..., m$ do STEPS 4 and 5.

    STEP 4 Set $x = a + (2i+1)h$.

    STEP 5 Set XI = XI $+ f(x)$.

STEP 6 Set XI = 2· XI·h.

16. a) $f^{(3)}$ is discontinuous at $x = 0.1$.

    b) 0.302506. Error bound 1.9×10^{-4}

    c) 0.302425. Actual value 0.302425.

18. 0.68271097

20. 9858 ft

22. 1054.694

24. a) no   b) 0.2792223   c) 0.2535482   d) 0.2798053   e) 0.2792222

    f) part e) gives best approximation; the error formulas could be applied only to e).

26. b) 0.6339100   c) 0.8886015   d) 3.142426

Exercise Set 4.5 (PAGE 190)

2. 14.02582, 29 nodes

4. 10.95015, 65 nodes

6. a) 0.3465748   b) −0.3465748

8. 1.145447

10. −0.023481944

Exercise Set 4.6 (Page 196)

2. a) 0.3465739 b) −0.3465739

4. $R_{6,6} = 0.9162907$

6. 14.02585

8. a) 0.7488276 b) 0.3024250

10. 0.2361

12.

$$R_{k,2} = \frac{4R_{k,1} - R_{k-1,1}}{3}$$

$$= \frac{1}{3}\left[R_{k-1,1} + 2h_{k-1}\sum_{i=1}^{2^{k-2}} f\left(a + \left(i - \frac{1}{2}\right)h_{k-1}\right)\right] \quad \text{from (4.62)}$$

$$= \frac{1}{3}\left[\frac{h_{k-1}}{2}(f(a) + f(b)) + h_{k-1}\sum_{i=1}^{2^{k-2}-1} f(a + ih_{k-1})\right.$$

$$\left. + 2h_{k-1}\sum_{i=1}^{2^{k-2}} f\left(a + \left(i - \frac{1}{2}\right)h_{k-1}\right)\right] \quad \text{from (4.61)}$$

$$= \frac{1}{3}\left[h_k(f(a) + f(b)) + 2h_k\sum_{i=1}^{2^{k-2}-1} f(a + 2ih_k) + 4h_k\sum_{i=1}^{2^{k-2}} f(a + (2i - 1)h_k)\right]$$

$$= \frac{h}{3}\left[f(a) + f(b) + 2\sum_{i=1}^{M-1} f(a + 2ih) + 4\sum_{i=1}^{M} f(a + (2i - 1)h)\right],$$

where $h = h_k$ and $M = 2^{k-2}$.

14.

$$R_{k,1} = \frac{h_k}{2}\left[f(a) + f(b) + 2\sum_{i=1}^{2^{k-1}-1} f(a + ih_k)\right]$$

$$= \frac{h_k}{2}\left[f(a) + f(b) + 2\sum_{i=1}^{2^{k-1}-1} f\left(a + \frac{i}{2}h_{k-1}\right)\right]$$

$$= \frac{h_k}{2}\left[f(a) + f(b) + 2\sum_{i=1}^{2^{k-2}-1} f(a + ih_{k-1}) + 2\sum_{i=1}^{2^{k-2}} f\left(a + \left(i - \frac{1}{2}\right)h_{k-1}\right)\right]$$

$$= \frac{1}{2}\left\{\frac{h_{k-1}}{2}\left[f(a) + f(b) + 2\sum_{i=1}^{2^{k-2}-1} f(a + ih_{k-1})\right] + h_{k-1}\sum_{i=1}^{2^{k-2}} f\left(a + \left(i - \frac{1}{2}\right)h_{k-1}\right)\right\}$$

$$= \frac{1}{2}\left[R_{k-1,1} + h_{k-1}\sum_{i=1}^{2^{k-2}} f\left(a + \left(i - \frac{1}{2}\right)h_{k-1}\right)\right].$$

Exercise Set 4.7 (PAGE 203)

2. a) 1.09804   b) 4.00000   c) 10.2106   d) 0.637062

   e) −5.52485   f) 0.718252

4. 11.14149, 10.94840, 10.95014

5. The Legendre polynomials P_2 and P_3 are given by

$$P_2(x) = \frac{(3x^2 - 1)}{2}$$

and

$$P_3(x) = \frac{(5x^3 - 3x)}{2},$$

so their roots are easily verified.

For $n = 2$,

$$c_1 = \int_{-1}^{1} \frac{x + 0.5773502692}{1.1547005}\, dx = 1$$

and

$$c_2 = \int_{-1}^{1} \frac{x - 0.5773502692}{-1.1547005}\, dx = 1.$$

For $n = 3$,

$$c_1 = \int_{-1}^{1} \frac{x(x + 0.7745966692)}{1.2}\, dx = \frac{5}{9},$$

$$c_2 = \int_{-1}^{1} \frac{(x + 0.7745966692)(x - 0.7745966692)}{-0.6}\, dx = \frac{8}{9},$$

and

$$c_3 = \int_{-1}^{1} \frac{x(x - 0.7745966692)}{1.2}\, dx = \frac{5}{9}.$$

6. The polynomial $L_n(x)$ has n distinct roots in $[0, \infty)$. (See Exercise 15 of Section 2.6.)
Let $x_1, ..., x_n$ be the n distinct roots in L_n. Define

$$c_i = \int_{0}^{\infty} e^{-x} \prod_{\substack{j=1 \\ j \neq i}}^{n} \frac{(x - x_j)}{(x_i - x_j)}\, dx$$

for each $i = 1, ..., n$.

Let P be any polynomial of degree $n - 1$ or less, and let P_{n-1} be the $(n - 1)$th Lagrange
polynomial for P on the nodes $x_1, ..., x_n$. As in the proof of Theorem 4.8,

$$\int_{0}^{\infty} P(x)e^{-x}\, dx = \int_{0}^{\infty} P_{n-1}(x)e^{-x}\, dx = \sum_{i=1}^{n} c_i P(x_i),$$

39

so the quadrature formula is exact for polynomials of degree $n - 1$ or less.

If P has degree $2n - 1$ or less, P can be divided by the nth Laguerre Polynomial L_n to obtain

$$P(x) = Q(x)L_n(x) + R(x),$$

where both Q and R are polynomials of degree less than n.

As in proof of Theorem 4.8, the orthogonality of the Laguerre polynomials on $[0, \infty)$ implies that

$$Q(x) = \sum_{i=0}^{n-1} d_i L_i(x)$$

for some constants d_i. Thus,

$$\int_0^\infty e^{-x} P(x)dx = \int_0^\infty \sum_{i=0}^{n-1} d_i L_i(x)L_n(x)e^{-x}dx + \int_0^\infty e^{-x} R(x)dx$$

$$= \sum_{i=0}^{n-1} d_i \int_0^\infty L_i(x)L_n(x)e^{-x}dx + \sum_{i=1}^{n} c_i R(x_i)$$

$$= 0 + \sum_{i=1}^{n} c_i R(x_i) = \sum_{i=1}^{n} c_i R(x_i).$$

But,

$$P(x_i) = Q(x_i)L_n(x_i) + R(x_i) = 0 + R(x_i) = R(x_i)$$

so

$$\int_0^\infty e^{-x} P(x)dx = \sum_{i=1}^{n} c_i P(x_i).$$

Hence the quadrature formula has degree of precision $2n - 1$.

8. 0.9238795, 0.9064405

10. If $\phi_k(x) < 0$ on (a, r_1), we can simply replace ϕ_k by $-\phi_k$ and obtain the same conclusion.

Exercise Set 4.8 (PAGE 215)

2. a) 2.945033   b) −20.14117

4. 13.15229

6. 1.469840

8. $\bar{x} = 0.3806333$, $\bar{y} = 0.3822558$

10. 1.031

12. 3.05200; Exact value 3.05213

14. a) $I_x = 29.066223$, $I_y = 29.097624$, $I_z = 13.427171$

b) $\hat{x} = 1.861632$, $\hat{y} = 1.862637$, $\hat{z} = 1.265295$

Chapter 5
Initial-Value Problems for Ordinary Differential Equations

Exercise Set 5.1 (PAGE 224)

2. a) Lipschitz constant $L = 1$; well-posed problem

b) Lipschitz constant $L = 1$; well-posed problem

c) Lipschitz constant $L = 1$; well-posed problem

d) f does not satisfy Lipschitz condition, so Theorem 5.6 cannot be used.

4. Let (t, y_1) and (t, y_2) be in D. Holding t fixed, define $g(y) = f(t, y)$. Suppose $y_1 < y_2$. Since the line joining (t, y_1) to (t, y_2) lies in D and f is continuous on D, we have $g \in C[y_1, y_2]$. Further, $g'(y) = \frac{\partial f(t, y)}{\partial y}$. Using the Mean Value Theorem on g, a number $\xi, \quad y_1 < \xi < y_2$, exists with

$$g(y_2) - g(y_1) = g'(\xi)(y_2 - y_1).$$

Thus,

$$f(t, y_2) - f(t, y_1) = \frac{\partial f(t, \xi)}{\partial y}(y_2 - y_1)$$

and

$$|f(t, y_2) - f(t, y_1)| \quad \leq \quad L|y_2 - y_1|.$$

The proof is similiar if $y_2 < y_1$. Thus, f satisfies a Lipschitz condition on D in the variable y with Lipschitz constant L.

6. Let the circle be described by

$$(t - t_0)^2 + (y - y_0)^2 = r^2$$

and let (t_1, y_1) and (t_2, y_2) be in the interior of the circle. Then

$$(t_1 - t_0)^2 + (y_1 - y_0)^2 < r^2$$

and

$$(t_2 - t_0)^2 + (y_2 - y_0)^2 < r^2.$$

Let λ be given, $\quad 0 < \lambda < 1$, and consider

$$(t_3, y_3) = \left(\lambda t_1 + (1 - \lambda)t_2, \lambda y_1 + (1 - \lambda)y_2 \right).$$

We first define $D = (t_3 - t_0)^2 + (y_3 - y_0)^2$. Then

$$\begin{aligned}
D &= (\lambda t_1 + (1 - \lambda)t_2 - t_0)^2 + (\lambda y_1 + (1 - \lambda)y_2 - y_0)^2 \\
&= (\lambda t_1 + (1 - \lambda)t_2 - \lambda t_0 - (1 - \lambda)t_0)^2 + (\lambda y_1 + (1 - \lambda)y_2 - \lambda y_0 - (1 - \lambda)y_0)^2 \\
&= [\lambda(t_1 - t_0) + (1 - \lambda)(t_2 - t_0)]^2 + [\lambda(y_1 - y_0) + (1 - \lambda)(y_2 - y_0)]^2 \\
&= \lambda^2(t_1 - t_0)^2 + 2\lambda(1 - \lambda)(t_1 - t_0)(t_2 - t_0) + (1 - \lambda)^2(t_2 - t_0)^2 \\
&\quad + \lambda^2(y_1 - y_0)^2 + 2\lambda(1 - \lambda)(y_1 - y_0)(y_2 - y_0) + (1 - \lambda)^2(y_2 - y_0)^2 \\
&\leq \lambda^2[(t_1 - t_0)^2 + (y_1 - y_0)^2] + (1 - \lambda)^2[(t_2 - t_0)^2 + (y_2 - y_0)^2] \\
&\quad + 2\lambda(1 - \lambda)[|t_1 - t_0||t_2 - t_0| + |y_1 - y_0||y_2 - y_0|]
\end{aligned}$$

But

$$\begin{aligned}
0 &\leq \Big(|t_2 - t_0||y_1 - y_0| - |y_2 - y_0||t_1 - t_0| \Big)^2 \\
&= |t_2 - t_0|^2|y_1 - y_0|^2 - 2|t_2 - t_0||y_1 - y_0||y_2 - y_0||t_1 - t_0| + |y_2 - y_0|^2|t_1 - t_0|^2,
\end{aligned}$$

so

$$2|t_2 - t_0||y_1 - y_0||y_2 - y_0||t_1 - t_0| \leq |t_2 - t_0|^2|y_1 - y_0|^2 + |y_2 - y_0|^2|t_1 - t_0|^2.$$

Thus,

$$\begin{aligned}
\Big(|t_1 - t_0||t_2 - t_0| + |y_1 - y_0||y_2 - y_0| \Big)^2 &= |t_1 - t_0|^2|t_2 - t_0|^2 \\
&\quad + 2|t_1 - t_0||t_2 - t_0||y_2 - y_0||y_1 - y_0| \\
&\quad + |y_1 - y_0|^2|y_2 - y_0|^2 \\
&\leq |t_1 - t_0|^2|t_2 - t_0|^2 + |t_2 - t_0|^2|y_1 - y_0|^2 \\
&\quad + |y_2 - y_0|^2|t_1 - t_0|^2 + |y_1 - y_0|^2|y_2 - y_0|^2 \\
&= \Big(|t_1 - t_0|^2 + |y_1 - y_0|^2 \Big)\big(|t_2 - t_0|^2 + |y_2 - y_0|^2\big).
\end{aligned}$$

Hence,

$$\begin{aligned}
D &\leq \lambda^2[(t_1 - t_0)^2 + (y_1 - y_0)^2] + (1 - \lambda)^2[(t_2 - t_0)^2 + (y_2 - y_0)^2] \\
&\quad + 2\lambda(1 - \lambda)[\sqrt{(t_1 - t_0)^2 + (y_1 - y_0)^2} \cdot \sqrt{(t_2 - t_0)^2 + (y_2 - y_0)^2}] \\
&< \lambda^2 r^2 + (1 - \lambda)^2 r^2 + 2\lambda(1 - \lambda)r \cdot r \\
&= [\lambda^2 + 2\lambda(1 - \lambda) + (1 - \lambda)^2]r^2 \\
&= (\lambda + (1 - \lambda))^2 r^2 = r^2.
\end{aligned}$$

Thus, (t_3, y_3) is in the interior of the circle and the circle is convex.

43

Exercise Set 5.2 (PAGE 232)

2. a)

t	w
0.5	0
1.0	0.125
1.5	0.625
2.0	1.75

b)

t	w
0.5	1
1.0	1.25
1.5	1.875
2.0	3.28125

c)

t	w
0.5	1
1.0	1.5
1.5	2.5
2.0	4

d)

t	w
1	4
2	0
3	0
4	0

4. a)

i	t_i	w_i
1	1.05	-0.9500000
2	1.10	-0.9045353
11	1.55	-0.6263495
12	1.60	-0.6049486
19	1.95	-0.4850416
20	2.00	-0.4712186

b) (i) -0.9481814 (ii) -0.6242094 (iii) -0.4773007

c) 0.029

6. a) $10^{-n/2}$

b) $10^{-n/2}(e-1) + 5e10^{-n-1}$

c)

t	$w(h = 0.1)$	$w(h = 0.01)$
0.5	0.40951	0.39499
1.0	0.65132	0.63397

8. As t increases the values quickly go to zero and are no longer adequate approximations. This behavior does not violate Theorem 5.9.

44

10.

j	t_j	w_j
20	2	0.702938
40	4	−0.0457793
60	6	0.294870
80	8	0.341673
100	10	0.139432

12. Using Eq. 5.21 we have

$$y_{i+1} - u_{i+1} = y_i - u_i + h[f(t_i, y_i) - f(t_i, u_i)] + \frac{h^2}{2}y''(\xi_i) - \delta_{i+1}.$$

Thus,

$$|y_{i+1} - u_{i+1}| \leq |y_i - u_i|(1 + hL) + \frac{h^2 M}{2} + \delta.$$

Using Lemma 5.8 with $a_j = |y_j - u_j|$ for each $j = 0, 1, ..., N$ while $s = hL$ and $t = \frac{h^2 M}{2} + \delta$, we have

$$|y_{i+1} - u_{i+1}| \leq e^{(i+1)hL}\left(|y_0 - u_0| + \frac{h^2 M}{2hL} + \frac{\delta}{hL}\right) - \frac{h^2 M}{2hL} - \frac{\delta}{hL}.$$

But $|y_0 - u_0| = |\delta_0|$, so

$$|y_{i+1} - u_{i+1}| \leq e^{(i+1)hL}\left(|\delta_0| + \frac{hM}{2L} + \frac{\delta}{hL}\right) - \frac{hM}{2L} - \frac{\delta}{hL}$$

from which (5.23) follows.

Exercise Set 5.3 (PAGE 238)

2. a)

i	t_i	w_i
1	1.1	1.215883
2	1.2	1.467561

b)

i	t_i	w_i
1	0.5	0.515625
2	1.0	1.09127

c)

i	t_i	w_i
1	1.5	-2
2	2	-1.679012
3	2.5	-1.484493
4	3	-1.374440

d)

i	t_i	w_i
1	0.25	1.086426
2	0.5	1.288245
3	0.75	1.512576
4	1	1.701494

4.

i	t_i	w_i
10	1.0	1.249305
20	2.0	2.095185

6. a)

i	t_i	w_i
1	1.05	-0.9525000
2	1.10	-0.9093138
11	1.55	-0.6459788
12	1.60	-0.6258649
19	1.95	-0.5139780
20	2.00	-0.5011957

c)

i	t_i	w_i
1	1.05	-0.9523813
2	1.10	-0.9090914
11	1.55	-0.6451629
12	1.60	-0.6250017
19	1.95	-0.5128226
20	2.00	-0.5000022

b) $y(1.052) \simeq -0.9507726$
$y(1.555) \simeq -0.6439674$
$y(1.978) \simeq -0.5068199$

d) $y(1.052) \simeq -0.9508220$
$y(1.555) \simeq -0.6457158$
$y(1.978) \simeq -0.5099844$

8. Yes, $\tau_{i+1} = \frac{h^2}{3} f''(t_i + \theta_i h, y(t_i + \theta_i h))$, where $0 < \theta_i < 1$ for each $i = 0, 1, \ldots, N-1$, provided $y \in C^3[a, b]$.

Exercise Set 5.4 (PAGE 247)

2. a)

i	t_i	w_i
1	1.1	1.214512
2	1.2	1.464112

b)

i	t_i	w_i
1	0.5	0.5131024
2	1.0	1.090105

c)

i	t_i	w_i
1	1.5	−1.700000
2	2.0	−1.484761
3	2.5	−1.362394
4	3.0	−1.287666

d)

i	t_i	w_i
1	0.25	1.093750
2	0.50	1.303161
3.	0.75	1.524441
4	1.0	1.706051

4.

i	t_i	Midpoint Method	Modified Euler	Heun's Method
10	1.0	1.251929	1.254358	1.252764
20	2.0	2.092963	2.090888	2.092247

6. a)

i	t_i	w_i
1	1.05	−0.9522677
2	1.10	−0.9088785
11	1.55	−0.6443787
12	1.60	−0.6241718
19	1.95	−0.5117111
20	2.00	−0.4988539

c)

i	t_i	w_i
1	1.05	−0.9523808
2	1.10	−0.9090909
11	1.55	−0.6451611
12	1.60	−0.6249998
19	1.95	−0.5128203
20	2.00	−0.4999998

b) $y(1.052) \simeq -0.9505321$, $y(1.555) \simeq -0.6423580$, $y(1.978) \simeq -0.5045111$

d) $y(1.052) \simeq -0.9505701$, $y(1.555) \simeq -0.6430856$,

8. a)

i	t_i	w_i
1	0.1	2.855833
4	0.4	1748.879
7	0.7	1073934
10	1.0	659529300

b)

i	t_i	w_i
4	0.1	0.03693409
16	0.4	0.1604301
28	0.7	0.4904167
40	1.0	1.000417

c)

i	t_i	w_i
10	0.1	0.01306468
40	0.4	0.1600333
70	0.7	0.4900333
100	1.0	1.000033

10. With $f(t,y) = -y + t + 1$, we have

$$w_i + hf(t_i + \frac{h}{2}, w_i + \frac{h}{2}f(t_i, w_i)) = w_i + \frac{h}{2}[f(t_i, w_i) + f(t_{i+1}, w_i + hf(t_i, w_i))]$$

$$= w_i + \frac{h}{4}\left[f(t_i, w_i) + 3f(t_i + \frac{2}{3}h, w_i + \frac{2}{3}hf(t_i, w_i))\right]$$

$$= w_i(1 - h + \frac{h^2}{2}) + t_i(h - \frac{h^2}{2}) + h.$$

12. 2099

14. a)

i	t_i	$v(t_i)$
0	0.0	8.000000
1	0.1	6.918721
2	0.2	5.864325
3	0.3	4.832189
4	0.4	3.818042
5	0.5	2.817886
6	0.6	1.827933
7	0.7	0.844542
8	0.8	−0.135820
9	0.9	−1.114963
10	1.0	−2.090146

b) $t = 0.8$

16. $\alpha_1 = \frac{1}{2}$, $\delta_1 = \frac{1}{2}$, $\alpha_2 = \frac{1}{2}$, $\delta_2 = \frac{1}{2}$, $\gamma_2 = \frac{1}{2}$, $\gamma_3 = \frac{1}{2}$, $\alpha_3 = 1$, $\delta_3 = 1$, $\gamma_4 = \frac{1}{2}$, $\gamma_5 = \frac{1}{2}$, $\gamma_6 = \frac{1}{2}$, $\gamma_7 = \frac{1}{2}$.

Exercise Set 5.5 (PAGE 255)

2. a)

i	t_i	w_i	h_i
1	0.13786398	0.1287829	0.13786398
2	0.27648532	0.2415554	0.13862133
3	0.41995798	0.3429257	0.14347266
4	0.56842575	0.4335838	0.14846777
5	0.72224714	0.5143406	0.15382139
6	0.88180938	0.5859672	0.15956225
7	1.00000000	0.6321208	0.11819062

b)

i	t_i	w_i
2	0.4	1.740635
8	1.6	2.003783
14	2.8	2.921607
20	4.0	4.036618

c)

i	t_i	w_i	h_i
1	1.14057827	0.5337650	0.14057827
2	1.27643648	1.4102643	0.13585821
3	1.41219779	2.7654863	0.13576130
4	1.54806082	4.7548100	0.13586303
5	1.68379803	7.5634068	0.13573721
6	1.81913323	11.4110813	0.13533520
7	1.95379561	16.5562736	0.13466238
8	2.00000000	18.6831545	0.04620439

d)

i	t_i	w_i
1	0.2	0.2027097
2	0.4	0.4227927
3	0.6	0.6841364
4	0.7839816	1.0000033

4. a) 80295.7   b) 80296

6. a) Change Algorithm 5.3 as follows:

STEP 3   Set $K_1 = hf(t, w); K_2 = hf(t + h, w + K_1)$.

STEP 4   Set $R = \frac{0.5|K_2 - K_1|}{h}$.

STEP 5   Set $\delta = \frac{0.5TOL}{R}$.

STEP 7   Set $t = t + h; w = w + K_1$.

49

b) (1a)

i	t_i	w_i
2	1.084453	1.173906
4	1.145802	1.315359
6	1.200000	1.451742

(1b)

i	t_i	w_i
2	0.5	0.5065512
4	1.0	1.066542

(1c)

i	t_i	w_i
6	1.088637	−1.845656
12	1.239981	−1.667065
21	2.068298	−1.293564
24	3.0	−1.166372

(1d)

i	t_i	w_i
3	0.5067633	0.0233816
10	1.027436	0.3116038
22	1.515484	1.085152
39	2.0	2.567287

Exercise Set 5.6 (PAGE 268)

2. All the starting values are from Algorithm 5.2.

a)

t_i	(5.63)	(5.64)	(5.65)	(5.66)
1.06	1.125548	1.125584	1.125585	1.125585
1.10	1.215801	1.215881	1.215884	1.215884
1.14	1.311752	1.311886	1.311891	1.311891
1.20	1.467316	1.467554	1.467563	1.467563

b)

t_i	(5.63)	(5.64)	(5.65)	(5.66)
0.4	0.4078158	0.4087538	0.4086147	0.4086183
0.6	0.6248659	0.6261072	0.6258409	0.6258518
0.8	0.8530337	0.8543221	0.8539445	0.8539629
1.0	1.091145	1.092261	1.091789	1.091815

c)

t_i	(5.63)	(5.64)	(5.65)	(5.66)
1.4	−1.608146	−1.555521	−1.555521	−1.555521
1.8	−1.429612	−1.359100	−1.404011	−1.384593
2.4	−1.296778	−1.239998	−1.278175	−1.246977
3.0	−1.225774	−1.182686	−1.212494	−1.192205

d)

t_i	(5.63)	(5.64)	(5.65)	(5.66)
0.4	1.207720	1.202474	1.200815	1.201486
0.6	1.389360	1.381041	1.379945	1.381294
0.8	1.562849	1.554372	1.554382	1.555453
1.0	1.707689	1.700923	1.701559	1.702134

4. a) g has a unique fixed point for

$$0 < h < \frac{24}{9}(\max_{0<t<0.2} e^{y(t)})^{-1}$$

b)

i	t_i	w_i
10	0.1	1.317218
20	0.2	1.784511

c) Newton's method will reduce the number of iterations per step from 4 to 3, using the stopping criterion

$$|w_i^{(k)} - w_i^{(k-1)}| \leq 10^{-6}.$$

6. Using the notation $y = y(t_i)$, $f = f(t_i, y(t_i))$, $f_t = f_t(t_i, y(t_i))$, etc., we have

$$y + hf + \frac{h^2}{2}(f_t + ff_y) + \frac{h^3}{6}(f_{tt} + f_tf_y + 2ff_{yt} + ff_y^2 + f^2f_{yy})$$

$$= y + ahf + bh[f - h(f_t + ff_y) + \frac{h^2}{2}(f_{tt} + f_tf_y + 2ff_{yt} + ff_y^2 + f^2f_{yy})]$$

$$+ ch[f - 2h(f_t + ff_y) + 2h^2(f_{tt} + f_tf_y + 2ff_{yt} + ff_y^2 + f^2f_{yy})].$$

Thus, $a + b + c = 1$, $-b - 2c = \frac{1}{2}$, and $\frac{1}{2}b + 2c = \frac{1}{6}$. This system has the solution $a = \frac{23}{12}$, $b = -\frac{16}{12}$, and $c = \frac{5}{12}$.

7. a) For some ξ_i in (t_{i-1}, t_i),

$$f(t, y(t)) = P_1(t) + \frac{f''(\xi_i, y(\xi_i))}{2}(t - t_i)(t - t_{i-1}).$$

Now

$$P_1(t) = \frac{(t - t_{i-1})}{(t_i - t_{i-1})}f(t_i, y(t_i)) + \frac{(t - t_i)}{(t_{i+1} - t_i)}f(t_{i-1}, y(t_{i-1})).$$

Thus,

$$\int_{t_i}^{t_{i+1}} P_1(t)dt = \frac{f(t_i, y(t_i))}{t_i - t_{i-1}} \int_{t_i}^{t_{i+1}} (t - t_{i-1})dt + \frac{f(t_{i-1}, y(t_{i-1}))}{t_{i-1} - t_i} \int_{t_i}^{t_{i+1}} (t - t_i)dt$$

$$= \frac{3h}{2}f(t_i, y(t_i)) - \frac{h}{2}f(t_{i-1}, y(t_{i-1})).$$

Further, $(t - t_i)(t - t_{i-1})$ does not change sign on (t_i, t_{i+1}) so the Mean Value Theorem for Integrals gives

$$\int_{t_i}^{t_{i+1}} \frac{f''(\xi_i, y(\xi_i))(t - t_i)(t - t_{i-1})}{2}dt = \frac{f''(\mu, y(\mu))}{2} \int_{t_i}^{t_{i+1}} (t - t_i)(t - t_{i-1})dt$$

$$= \frac{-5h^2f''(\mu, y(\mu))}{12}.$$

Replacing $y(t_j)$ with w_j for $j = i - 1, i, i + 1$ in the formula

$$y(t_{i+1}) = y(t_i) + \int_{t_i}^{t_{i+1}} f(t, y(t))dt$$

gives

$$w_{i+1} = w_i + \frac{h[3f(t_i, w_i) - f(t_{i-1}, w_{i-1})]}{2}$$

and the local truncation error is

$$\tau_{i+1}(h) = \frac{-5h^2y'''(\mu)}{12}.$$

52

b) Using the backward difference polynomial with $m = 4$ gives

$$\int_{t_i}^{t_{i+1}} f(t, y(t))dt = \sum_{k=0}^{3} \nabla^k f(t_i, y(t_i))h(-1)^k \int_0^1 \binom{-s}{k} ds$$
$$+ \frac{h^5}{24} \int_0^1 s(s+1)(s+2)(s+3)f^{(4)}(\xi_i, y(\xi_i))ds.$$

From Table 5.9,

$$\int_{t_i}^{t_{i+1}} f(t, y(t))dt = h[f(t_i, y(t_i)) + \frac{1}{2}\nabla f(t_i, y(t_i)) + \frac{5}{12}\nabla^2 f(t_i, y(t_i))$$
$$+ \frac{3}{8}\nabla^3 f(t_i, y(t_i))]$$
$$+ \frac{h^5}{24} \int_0^1 s(s+1)(s+2)(s+3)f^{(4)}(\xi_i, y(\xi_i))ds.$$

Since

$$\nabla f(t_i, y(t_i)) = f(t_i, y(t_i)) - f(t_{i-1}, y(t_{i-1})),$$
$$\nabla^2 f(t_i, y(t_i)) = f(t_i, y(t_i)) - 2f(t_{i-1}, y(t_{i-1})) + f(t_{i-2}, y(t_{i-2})),$$
$$\nabla^3 f(t_i, y(t_i)) = f(t_i, y(t_i)) - 3f(t_{i-1}, y(t_{i-1})) + 3f(t_{i-2}, y(t_{i-2}))$$
$$+ f(t_{i-3}, y(t_{i-3})),$$

and $s(s+1)(s+2)(s+3)$ does not change sign on $(0,1)$, we can simplify and use the Mean Value Theorem for integrals to obtain

$$\int_{t_i}^{t_{i+1}} f(t, y(t))dt = h\left[55f(t_i, y(t_i)) - 59f(t_{i-1}, y(t_{i-1})) + 37f(t_{i-2}, y(t_{i-2}))\right.$$
$$\left. - 9f(t_{i-3}, y(t_{i-3}))\right] + \frac{h^5}{24}f^{(4)}(\mu, y(\mu)) \int_0^1 s(s+1)(s+2)(s+3)ds$$

for some μ in (t_{i-3}, t_{i+1}). Since

$$\int_0^1 s(s+1)(s+2)(s+3)ds = \frac{251}{30},$$

(5.65) follows.

8.

$$f(t, y(t)) = \frac{1}{2h^2}(t - t_i)(t - t_{i+1})f(t_{i-1}, y(t_{i-1})) - \frac{1}{h^2}(t - t_{i-1})(t - t_{i+1})f(t_i, y(t_i))$$
$$+ \frac{1}{2h^2}(t - t_{i-1})(t - t_i)f(t_{i+1}, y(t_{i+1}))$$
$$+ \frac{1}{6}\frac{d^3}{dt^3}f(\xi, y(\xi))(t - t_{i-1})(t - t_i)(t - t_{i+1}).$$

53

So

$$\int_{t_i}^{t_{i+1}} y'(t)\, dt = \int_{t_i}^{t_{i+1}} f(t, y(t))\, dt$$

and

$$
\begin{aligned}
y(t_{i+1}) - y(t_i) =& \frac{f(t_{i-1}, y(t_{i-1}))}{2h^2} \int_{t_i}^{t_{i+1}} (t - t_i)(t - t_{i+1})dt \\
& - \frac{f(t_i, y(t_i))}{h^2} \int_{t_i}^{t_{i+1}} (t - t_{i-1})(t - t_i)dt \\
& + \frac{f(t_{i+1}, y(t_{i+1}))}{2h^2} \int_{t_i}^{t_{i+1}} (t - t_{i-1})(t - t_i)dt \\
& + \int_{t_i}^{t_{i+1}} \frac{f'''(\xi, y(\xi))}{6}(t - t_{i-1})(t - t_i)(t - t_{i+1})dt \\
=& \frac{-h}{12} f(t_{i-1}, y(t_{i-1})) + \frac{2h}{3} f(t_i, y(t_i)) + \frac{5h}{12} f(t_{i+1}, y(t_{i+1})) \\
& + \frac{f'''(\mu, y(\mu))}{6} \int_{t_i}^{t_{i+1}} (t - t_{i-1})(t - t_i)(t - t_{i+1})dt.
\end{aligned}
$$

The last part follows from Theorem 4.2. Further integration yields formula (5.67) and the local truncation error.

10.

$$
\begin{aligned}
y(t_{i+1}) - y(t_{i-1}) =& \int_{t_{i-1}}^{t_{i+1}} f(t, y(t))dt \\
=& \frac{h}{3}[f(t_{i-1}, y(t_{i-1})) + 4f(t_i, y(t_i)) + f(t_{i+1}, y(t_{i+1}))] \\
& - \frac{h^5}{90} f^{(4)}(\xi, y(\xi)).
\end{aligned}
$$

This leads to the difference equation

$$w_{i+1} = w_{i-1} + \frac{h[f(t_{i-1}, w_{i-1}) + 4f(t_i, w_i) + f(t_{i+1}, w_{i+1})]}{3}$$

with local truncation error

$$\tau_{i+1}(h) = \frac{-h^4 y^{(5)}(\xi)}{90}.$$

11. For Simpson's method refer to Exercise 10. To derive Milne's method, integrate $y'(t) = f(t, y(t))$ over the interval $[t_{i-3}, t_{i+1}]$ to obtain

$$y(t_{i+1}) - y(t_{i-3}) = \int_{t_{i-3}}^{t_{i+1}} f(t, y(t))dt.$$

54

Using the open Newton-Cotes formula (4.48), we have

$$y(t_{i+1}) - y(t_{i-3}) = \frac{4h[2f(t_i, y(t_i)) - f(t_{i-1}, y(t_{i-1})) + 2f(t_{i-2}, y(t_{i-2}))]}{3}$$
$$+ \frac{14h^5 f^{(4)}(\xi, y(\xi))}{45}.$$

The difference equation becomes

$$w_{i+1} = w_{i-3} + \frac{h[8f(t_i, w_i) - 4f(t_{i-1}, w_{i-1}) + 8f(t_{i-2}, w_{i-2})]}{3}$$

with local truncation error

$$\tau_{i+1}(h) = \frac{14h^5 y^{(5)}(\xi)}{45}.$$

12.

$$k = 0 : (-1)^k \int_0^1 \binom{-s}{k} ds = \int_0^1 ds = 1$$

$$k = 1 : (-1)^k \int_0^1 \binom{-s}{k} ds = -\int_0^1 -s\, ds = \frac{1}{2}$$

$$k = 2 : (-1)^k \int_0^1 \binom{-s}{k} ds = \int_0^1 \frac{s(s+1)}{2} ds = \frac{5}{12}$$

$$k = 3 : (-1)^k \int_0^1 \binom{-s}{k} ds = -\int_0^1 \frac{-s(s+1)(s+2)}{6} ds = \frac{3}{8}$$

$$k = 4 : (-1)^k \int_0^1 \binom{-s}{k} ds = \int_0^1 \frac{s(s+1)(s+2)(s+3)}{24} ds = \frac{251}{720}$$

$$k = 5 : (-1)^k \int_0^1 \binom{-s}{k} ds = -\int_0^1 -\frac{s(s+1)(s+2)(s+3)(s+4)}{120} ds = \frac{95}{288}$$

Exercise Set 5.7 (PAGE 274)

2. a)

i	t_i	w_i
3	0.28636288	0.24900997
8	0.57272576	0.43601413
12	0.85908864	0.57645235
17	1.00000000	0.63212081

b)

i	t_i	w_i
6	1.0	1.735730
13	2.4	2.581396
20	3.8	3.844719
26	5.0	5.013462

c)

i	t_i	w_i
5	1.244171	1.163733
9	1.488342	3.791232
13	1.732513	8.814560
21	2.000000	18.68314

d)

i	t_i	w_i
4	0.2329400	0.2372463
7	0.4658800	0.5027982
10	0.6988200	0.8402934
14	0.8647846	1.172891

4. 8.693298

5. Modify algorithm 5.5 as follows:

delete STEP 12

change STEP 13 Set $h = 2h$

delete STEP 17

change STEP 18 Set $h = 0.5h$

One could also use previous values when doubling the stepsize by setting

$$w_{i-2} = w_{i-3}, \ t_{i-2} = t_{i-3}; \quad w_{i-3} = w_{i-5}, \ t_{i-3} = t_{i-5}; \quad \text{and} \quad w_{i-4} = w_{i-7}, \ t_{i-4} = t_{i-7}.$$

6.

(2a)

i	t_i	w_i	h
5	0.25	0.2211992	0.05
10	0.5	0.3934694	0.05
17	1.0	0.6321206	0.0875

(2b)

i	t_i	w_i	h
5	1.0	1.735731	0.2
10	2.0	2.270629	0.2
16	3.0	3.099550	0.1
26	4.0	4.036623	0.1
34	5.0	5.013473	0.15

(2c)

i	t_i	w_i	h
5	1.25	1.206345	0.05
10	1.50	3.967670	0.05
15	1.75	9.298744	0.05
23	2.0	18.68312	0.0125

(2d)

i	t_i	w_i	h
3	0.3	0.3093360	0.1
7	0.6	0.6841509	0.05
10	0.75	0.9316168	0.05
14	0.7853982	1.000022	0.008849541

Exercise Set 5.8 (PAGE 281)

2. a)

i	t_i	w_i
1	0.2	0.1812692
3	0.6	0.4511883
5	1.0	0.63211205

c)

i	t_i	w_i
2	1.4	2.620056
4	1.8	10.79313
5	2.0	18.68248

b)

i	t_i	w_i
5	1.0	1.735724
12	2.4	2.581427
19	3.8	3.844742
25	5.0	5.013477

d)

i	t_i	w_i
1	0.2	0.2027100
2	0.4	0.4227967
3	0.6	0.6841405
4	0.78539816	1.000005

4. The error term is

$$\frac{h_0 O(h_1^2) - h_1 O(h_0^2)}{h_0 - h_1} = \frac{h_0}{h_0 - h_1} O(h_1^2) - \frac{h_1}{h_0 - h_1} O(h_0^2).$$

Since $h_1 < h_0$, we have $K h_1^2 < K h_0^2$ for any constant K.
Thus,

$$\frac{h_0 O(h_1^2) - h_1 O(h_0^2)}{h_0 - h_1} = \frac{h_0}{h_0 - h_1} O(h_1^2) - \frac{h_1}{h_0 - h_1} O(h_0^2) = \frac{h_0 - h_1}{h_0 - h_1} O(h_0^2) = O(h_0^2).$$

Exercise Set 5.9 (PAGE 289)

2. a)

i	t_i	w_{1i}	w_{2i}
5	0.5	0.71544570	−0.19435011
10	1.0	0.77153962	0.40366236
15	1.5	1.1204202	1.0088608
20	2.0	1.8134261	1.8134371

b)

i	t_i	w_{1i}	w_{2i}
5	1.25	0.9405698	-0.4613824
10	1.50	0.7779721	-0.8190555
15	1.75	0.5425639	-1.038640
20	2.00	0.2725883	-1.091117

c)

i	t_i	w_{1i}	w_{2i}	w_{3i}
5	1.0	3.731627	4.181249	4.457219
10	2.0	11.31425	12.50243	13.75296
15	3.0	34.04396	37.36869	40.73623

d)

i	t_i	w_{1i}	w_{2i}	w_{3i}
2	1.1	-1.111110	-1.234565	-2.743482
6	1.3	-1.428566	-2.040798	-5.830883
12	1.6	-2.499944	-6.249534	-31.24937
18	1.9	-9.967667	-98.91652	-1990.752

4. (1a)

i	t_i	w_{1i}	w_{2i}	h_i
4	0.3033254	1.272848	2.011206	0.06607599
8	0.5169403	4.221085	4.817428	0.04749020
14	0.7498743	14.00761	14.48004	0.03396841
23	1.0000000	49.34951	49.71739	0.02326297

i	t_i	w_{1i}	w_{2i}	h_i
5	0.5	0.9567304	-1.083835	0.1
10	1.0	1.306562	-0.8329700	0.1
15	1.5	1.344183	-0.5698183	0.1
20	2.0	1.143339	-0.3693763	0.1

(1c)

i	t_i	w_{1i}	w_{2i}	w_{3i}	h_i
2	0.2	2.997467	-0.0373335	0.7813972	0.1
5	0.5	2.963539	-0.2083737	0.3981570	0.1
7	0.7	2.905645	-0.3759609	0.1206244	0.1
10	1.0	2.749653	-0.6690483	-0.3011688	0.1

(1d)

i	t_i	w_{1i}	w_{2i}	w_{3i}	h_i
2	0.2	1.381653	1.008000	-0.6183308	0.1
5	0.5	1.907532	1.125000	-0.0909057	0.1
7	0.7	2.255036	1.343000	0.2634397	0.1
10	1.0	2.832121	2.000000	0.8821206	0.1

6.

i	t_i	x_{1i}	x_{2i}
2	1.00000000	8716	1435
4	2.00000000	7907	2120
6	3.00000000	6666	2813

Exercise Set 5.10 (PAGE 300)

2. For the Adams-Bashforth Method,

$$F(t_i, h, w_{i+1}, w_i, w_{i-1}, w_{i-2}, w_{i-3}) = \frac{1}{24}\left[55f(t_i, w_i) - 59f(t_{i-1}, w_{i-1}) \right.$$

$$\left. + 37f(t_{i-2}, w_{i-2}) - 9f(t_{i-3}, w_{i-3}) \right],$$

so if $f \equiv 0$, then $F \equiv 0$. If f has Lipschitz constant L, then

$$|F(t_i, h, w_{i+1}, \ldots, w_{i-3}) - F(t_i, h, v_{i+1}, \ldots, v_{i-3})| \leq \frac{L}{24} \Big[55|w_i - v_i| + 59|w_{i-1} - v_{i-1}|$$

$$+ 37|w_{i-2} - v_{i-2}| + 9|w_{i-3} - v_{i-3}| \Big],$$

so $C = \frac{59L}{24}$ will suffice. A similar result holds for the Adams-Moulton method, but with $C = \frac{19L}{24}$.

4. b) $w_2 = 0.18065 \approx y(0.2), w_5 = 0.35785 \approx y(0.5), w_7 = 0.15342 \approx y(0.7)$, and

$w_{10} = -9.7414 \approx y(1.0)$.

c) $w_{20} = -60.402 \simeq y(0.2), w_{50} = -1.37 \times 10^{17} \simeq y(0.5), w_{70} = -5.11 \times 10^{26} \simeq y(0.7)$,

and $w_{100} = -1.16 \times 10^{41} \simeq y(1.0)$.

6. For $h = 0.1$:

$w_{10} = 0.367883 \simeq y(1) = 0.3678794$, and $w_{100} = 3.84917 \simeq y(10) = 0.0000454$.

For $h = 0.01$:

$w_{100} = 0.367879 \simeq y(1) = 0.3678794$ and $w_{1000} = 0.000109962 \simeq y(10) = 0.0000454$.

8. $4\varepsilon, 13\varepsilon, 40\varepsilon, 121\varepsilon, 364\varepsilon$

10. a) $p(\beta) = (1 - \frac{h\lambda}{3})\beta^2 - \frac{4h\lambda}{3}\beta - (1 - \frac{h\lambda}{3})$

b) If $hL < 0$, there is a root $\beta_i < -1$.

Exercise Set 5.11 (Page 309)

2. 1(a)

t_i	Algorithm 5.4	t_i	Algorithm 5.5
0.2	1.000793	0.1887722	1.057700
0.5	0.2199580	0.5191235	0.2026877
0.7	0.07986897	0.6856033	0.08816528
1.0	0.01741607	0.9739779	0.02081630

1(b)

t_i	Algorithm 5.4	t_i	Algorithm 5.5
0.2	0.9435276	0.1900519	0.9831462
0.5	0.5750137	0.5167139	0.5972784
0.7	0.7165464	0.7031522	0.7249891
1.0	1.001479	0.9576240	0.9612774

6. a) The Trapezoidal method applied to the test equation gives

$$w_{j+1} = \frac{1 + \frac{h\lambda}{2}}{1 - \frac{h\lambda}{2}}\, w_j$$

so

$$Q(h\lambda) = \frac{2 + h\lambda}{2 - h\lambda}.$$

Thus, $|Q(h\lambda)| < 1$ whenever $\text{Re}(h\lambda) < 0$.

b) The Backward Euler method applied to the test equation gives

$$w_{j+1} = \frac{w_j}{1 - h\lambda}$$

so

$$Q(h\lambda) = \frac{1}{1 - h\lambda}.$$

Thus, $|Q(h\lambda)| < 1$ whenever $\text{Re}(h\lambda) < 0$.

Chapter 6
Direct Methods for Solving Linear Systems

Exercise Set 6.1 (PAGE 316)

2. a) $x_1 = 1$, $x_2 = 1$ b) No solutions.

    c) An infinite number of solutions.   d) $x_1 = 5$, $x_2 = 3$

4. a) Represents 3 lines in the plane that do not intersect in a common point.

    b) Represents 2 planes in space that intersect in a line.

6. Suppose $x'_1, ..., x'_n$ is a solution to the linear system (6.1).

    (i) The new system becomes

$$E_1 : a_{11}x_1 + a_{12}x_2 + ... + a_{1n}x_n = b_1$$

$$\vdots$$

$$E_i : \lambda a_{i1}x_1 + \lambda a_{i2}x_2 + ... + \lambda a_{in}x_n = \lambda b_i$$

$$\vdots$$

$$E_n : a_{n1}x_1 + a_{n2}x_2 + ... + a_{nn}x_n = b_n.$$

Clearly, $x'_1, ..., x'_n$ satisfies this system.

Conversely, if $x^*_1, ..., x^*_n$ satisfies the new system, dividing E_i by λ shows $x^*_1, ..., x^*_n$ satisfies (6.1).

(ii) The new system becomes

$$E_1 : a_{11}x_1 + a_{12}x_2 + ... + a_{1n}x_n = b_1$$

$$\vdots$$

$$E_i : (a_{i1} + \lambda a_{j1})x_1 + (a_{i2} + \lambda a_{j2})x_2 + ... + (a_{in} + \lambda a_{jn})x_n = b_i + \lambda b_j$$

$$\vdots$$

$$E_n : a_{n1}x_1 + a_{n2}x_2 + ... + a_{nn}x_n = b_n.$$

Clearly, $x'_1, ..., x'_n$ satisfies all but possibly the ith equation. Multiplying E_j by λ yields

$$\lambda a_{j1}x'_1 + \lambda a_{j2}x'_2 + ... + \lambda a_{jn}x'_n = \lambda b_j,$$

so simplifying E_i in the new system results in the system (6.1). Thus, $x'_1, ..., x'_n$ satisfies the new system.

Conversely, if $x_1^*, ..., x_n^*$ is a solution to the new system, then all but possibly E_i of (6.1) are satisfied by $x_1^*, ..., x_n^*$. Multiplying E_j of the new system by $-\lambda$ gives

$$-\lambda a_{j1} x_1^* - \lambda a_{j2} x_2^* - ... - \lambda a_{jn} x_n^* = -\lambda b_j.$$

Subtracting this from E_i in the new system yields E_i of (6.1). Thus, $x_1^*, ..., x_n^*$ is a solution of (6.1).

(iii) The new system and the old system have the same set of equations to satisfy. Thus, they have the same solution set.

Exercise Set 6.2 (PAGE 325)

2. a) $x_1 = 1.1875$, $x_2 = 1.8125$, $x_3 = 0.875$; one row interchange is necessary.

   b) $x_1 = 0.75$, $x_2 = 0.5$, $x_3 = -0.125$; one row interchange is necessary.

   c) $x_1 = -1$, $x_2 = 0$, $x_3 = 1$; no interchanges are required.

   d) $x_1 = 1$, $x_2 = 2$, $x_3 = -1$; no interchanges are required.

   e) $x_1 = 1.5$, $x_2 = 2$, $x_3 = -1.2$ $x_4 = 3$; no interchanges are required.

   f) $x_1 = \frac{22}{9}$, $x_2 = -\frac{4}{9}$, $x_3 = \frac{4}{3}$, $x_4 = 1$; one row interchange is necessary.

4. a) $\alpha = -1/3$

   b) $\alpha = 1/3$, $x_1 = x_2 + 1.5$, x_2 arbitrary

   c) If $\alpha \neq \pm 1/3$,
   $$x_1 = \frac{3}{2(1 + 3\alpha)} \quad \text{and} \quad x_2 = \frac{-3}{2(1 + 3\alpha)}.$$

6. Change Algorithm 6.1 as follows:

   STEP 1 For $i = 1, ... , n$ do STEPS 2, 3, and 4.

   STEP 4 For $j = 1, ... , i - 1, i + 1, ... , n$ do STEPS 5 and 6.

   STEP 8 For $i = 1, ... , n$ set $x_i = a_{i,n+1}/a_{ii}$.

   In addition, delete STEP 9.

8. a) -227.0787, 476.9262, -177.6934

   b) 1.000036, 0.9999991, 0.9986052

   c) 0.9999997, 1.000001, 0.9999994, 0.9999995

   d) -0.03177120, 0.5955572, -2.381768, 2.778329

9. a) Referring to Exercise 6 gives the algorithm for the Gauss-Jordan method. The following operation counts are derived:

	Multiplications/Divisions	Additions/Subtractions
STEP 5	$n-1$ for each i	0
STEP 6	$(n-1)(n-i+1)$ for each i	$(n-1)(n-i+1)$ for each i
STEP 8	1 for each i	0

Thus, the totals are

Multiplications/Divisions: $\displaystyle\sum_{i=1}^{n}[(n-1)+(n-1)(n-i+1)+1]=\frac{n^3}{2}+n^2-\frac{n}{2}.$

Additions/Subtractions: $\displaystyle\sum_{i=1}^{n}(n-1)(n-i+1)=\frac{n^3-n}{2}.$

10. a) The Gaussian elimination procedure requires

$$\frac{(2n^3+3n^2-5n)}{6} \text{ Multiplications/Divisions}$$

and

$$\frac{n^3-n}{3} \text{ Additions/Subtractions.}$$

The additional elimination steps are:

For $i=n, n-1, ..., 2$

for $j=1, ..., i-1,$

$$\text{set}\quad a_{j,n+1}=\frac{a_{j,n+1}-a_{ji}a_{i,n+1}}{a_{ii}}.$$

This requires

$$n(n-1) \text{ Multiplications/Divisions}$$

and

$$\frac{n(n-1)}{2} \text{ Additions/Subtractions.}$$

Solving for

$$x_i=\frac{a_{i,n+1}}{a_{ii}}$$

requires n divisions. Thus, the totals are

$$\frac{n^3}{3}+\frac{3n^2}{2}-\frac{5n}{6} \text{ Multiplications/Divisions}$$

and

$$\frac{n^3}{3}+\frac{n^2}{2}-\frac{5n}{6} \text{ Additions/Subtractions.}$$

9. b) and **10.** b). The results for these exercises are listed in the following table. In this table the abbreviations M/D and A/S are used for Multiplications/Divisions and for Additions/Subtractions, respectively.

	Gaussian Elimination		Gauss-Jordan		Hybrid	
n	M/D	A/S	M/D	A/S	M/D	A/S
3	17	11	21	12	20	11
10	430	375	595	495	475	375
50	44150	42875	64975	62475	45375	42875
100	343300	338250	509950	499950	348250	338250

12. a) -227.0788, 476.9262, -177.6934

    b) 0.9990999, 0.9999991, 0.9986052

    c) 1.000000, 1.000001, 0.9999994, 0.9999995

    d) -0.03177060, 0.5955554, -2.381768, 2.778329

Exercise Set 6.3 (PAGE 334)

2. a) $x_1 = -10.0$, $x_2 = 1.01$; $x_1 = 10.0$, $x_2 = 1.00$; $x_1 = 10.0$, $x_2 = 1.00$.

    b) $x_1 = 1.00$, $x_2 = 10.0$; $x_1 = 1.00$, $x_2 = 10.0$; $x_1 = 1.00$, $x_2 = 10.0$.

    c) $x_1 = -8.80$, $x_2 = 39.2$, $x_3 = -30.3$;

        $x_1 = -9.85$, $x_2 = 45.2$, $x_3 = -36.1$;

        $x_1 = -8.80$, $x_2 = 39.2$, $x_3 = -30.3$.

    d) $(0.102, 1.00, 2.42)$; $(-0.113, 1.40, 2.42)$; $(-0.113, 1.40, 2.42)$.

4. a) $x_1 = 0.9260$, $x_2 = 14.29$; $x_1 = 0.8275$, $x_2 = 20.00$; $x_1 = 0.9260$, $x_2 = 14.29$.

    b) $x_1 = -10.00$, $x_2 = 1.000$; $x_1 = 10.00$, $x_2 = 1.000$; $x_1 = 10.00$, $x_2 = 1.000$.

    c) $x_1 = 1.196$, $x_2 = 1.152$; $x_1 = 1.153$, $x_2 = 1.153$; $x_1 = 1.153$, $x_2 = 1.153$.

    d) $x_1 = 1.153$, $x_2 = 1.153$; $x_1 = 1.153$, $x_2 = 1.153$; $x_1 = 1.153$, $x_2 = 1.153$.

6. Change Algorithm 6.2 as follows:

Add to STEP 1.

        $NCOL(i) = i$

Replace STEP 3 with the following.

    Let p and q be the smallest integers with $i \le p$, $q \le n$ and

$$|a(NROW(p), NCOL(q))| = \max_{i \le k, j \le n} |a(NROW(k), NCOL(j))|.$$

Add to STEP 4.

        $A(NROW(p), NCOL(q)) = 0$

Add to STEP 5.

If $NCOL(q) \neq NCOL(i)$ then

set

$$NCOPY = NCOL(i);$$
$$NCOL(i) = NCOL(q);$$
$$NCOL(q) = NCOPY.$$

Replace STEP 7 with the following.

Set

$$m(NROW(j), NCOL(i)) = \frac{a(NROW(j), NCOL(i))}{a(NROW(i), NCOL(i))}.$$

Replace in STEP 8:

$m(NROW(j), i)$ by $m(NROW(j), NCOL(i))$

Replace in STEP 9:

$a(NROW(n), n)$ by $a(NROW(n), NCOL(n))$

Replace STEP 10 with the following.

Set

$$X(NCOL(n)) = \frac{a(NROW(n), n+1)}{a(NROW(n), NCOL(n))}.$$

Replace STEP 11 with the following.

Set

$$X(NCOL(i)) = \frac{a(NROW(i), n+1) - \sum_{j=i+1}^{n} a(NROW(i), NCOL(j)) \cdot X(NCOL(j))}{A(NROW(i), NCOL(i))}.$$

Replace STEP 12 with the following.

OUTPUT $(X(NCOL(i))$ for $i = 1, \ldots, n)$.

7. b) 2a) 10.0, 1.00   2b) 1.0, 10.0
   2c) −10.0, 45.4, −36.4   2d) −0.113, 1.40, 2.42
   c) 3a) 9.98, 1.00   3b) 1.0, 9.98
   3c) −9.66, 45.0, −36.0   3d) −0.102, 1.38, 2.42
   d) 4a) 0.8275, 20.00   4b) 0.8275, 20.00
   4c) 1.153, 1.153   4d) 1.153, 1.153

Exercise Set 6.4 (PAGE 344)

2. a) $\begin{bmatrix} 1 & 0 & 0 \\ 1 & 2 & 0 \\ 9 & 5 & 1 \end{bmatrix}$ **b)** $\begin{bmatrix} 1 & -1 & 2 \\ 2 & -1 & 7 \\ -2 & 1 & -5 \end{bmatrix}$ **c)** $\begin{bmatrix} 1 & 0 & 0 \\ 2 & 1 & 0 \\ -7 & -2 & 1 \end{bmatrix}$ **d)** $\begin{bmatrix} 6 & -7 & 15 \\ 0 & -1 & 3 \\ 0 & 0 & 6 \end{bmatrix}$

3. a) (a)

$$(A^t)^t = \left(\begin{bmatrix} 4 & 6 & 1 & -1 \\ 2 & 1 & 0 & 0.5 \\ 3 & 0 & 0 & 1 \\ 1 & -1 & 1 & 1 \end{bmatrix}^t \right)^t = \begin{bmatrix} 4 & 2 & 3 & 1 \\ 6 & 1 & 0 & -1 \\ 1 & 0 & 0 & 1 \\ -1 & 0.5 & 1 & 1 \end{bmatrix}^t = \begin{bmatrix} 4 & 6 & 1 & -1 \\ 2 & 1 & 0 & 0.5 \\ 3 & 0 & 0 & 1 \\ 1 & -1 & 1 & 1 \end{bmatrix} = A$$

(b)

$$(A+B)^t = \begin{bmatrix} 5 & 8 & 4 & 3 \\ 2 & 3 & -1 & 1.5 \\ 3 & 0 & 3 & 3 \\ 1 & -1 & 1 & 0 \end{bmatrix}^t = \begin{bmatrix} 5 & 2 & 3 & 1 \\ 8 & 3 & 0 & -1 \\ 4 & -1 & 3 & 1 \\ 3 & 1.5 & 3 & 0 \end{bmatrix} = A^t + B^t$$

(c)

$$(AB)^t = \begin{bmatrix} 4 & 20 & 9 & 25 \\ 2 & 6 & 5 & 8.5 \\ 3 & 6 & 9 & 11 \\ 1 & 0 & 7 & 4 \end{bmatrix}^t = \begin{bmatrix} 4 & 2 & 3 & 1 \\ 20 & 6 & 6 & 0 \\ 9 & 5 & 9 & 7 \\ 25 & 8.5 & 11 & 4 \end{bmatrix} = B^t A^t$$

(d)

$$(A^{-1})^t = \begin{bmatrix} \frac{2}{11} & \frac{-1}{11} & \frac{3}{11} & \frac{-6}{11} \\ \frac{-14}{11} & \frac{18}{11} & \frac{-10}{11} & \frac{42}{11} \\ 1 & -1 & 0 & -2 \\ \frac{-2}{11} & \frac{1}{11} & \frac{8}{11} & \frac{6}{11} \end{bmatrix} = (A^t)^{-1}$$

b) (a) $[(A^t)^t]_{i,j} = (A^t)_{j,i} = A_{ij}$ for each i and j.

(b) $[(A+B)^t]_{i,j} = (A+B)_{ji} = a_{ji} + b_{ji}$ and $(A^t)_{ij} + (B^t)_{ij} = A_{ji} + B_{ji} = a_{ji} + b_{ji}$.

(c) Suppose A is an $m \times n$ matrix and B is an $n \times p$ matrix. Then for $1 \le i \le p$ and $1 \le j \le m$,

$$[(AB)^t]_{ij} = (AB)_{j,i} = \sum_{k=1}^{n} a_{jk} b_{ki}.$$

Note that B^t is an $p \times n$ matrix and A^t is an $n \times m$ matrix, so

$$(B^t A^t)_{ij} = \sum_{k=1}^{n} B^t_{ik} A^t_{kj} = \sum_{k=1}^{n} b_{ki} a_{jk}$$

67

for $1 \leq i \leq p$ and $1 \leq j \leq m$.

(d) Suppose A^{-1} exists. Then

$$A^t(A^{-1})^t = (A^{-1}A)^t = I$$

by part (c). Similiarly,

$$(A^{-1})^t A^t = I \quad \text{so} \quad (A^t)^{-1} = (A^{-1})^t.$$

4. a) Let

$$A = \begin{bmatrix} 2 & 1 \\ 1 & 0 \end{bmatrix} \quad \text{and} \quad B = \begin{bmatrix} 1 & -1 \\ -1 & 2 \end{bmatrix}.$$

Then

$$AB = \begin{bmatrix} 1 & 0 \\ 1 & -1 \end{bmatrix}$$

is not symmetric.

b) Let A be a nonsingular symmetric matrix. By Theorem 6.12 (d), $(A^{-1})^t = (A^t)^{-1}$. Thus, $(A^{-1})^t = (A^t)^{-1} = A^{-1}$ and A^{-1} is symmetric.

c) Not true. Use matrices A and B from part (a).

6. Parts a), b), and c):

The first system has $x_1 = 3, x_2 = -6, x_3 = -2, x_4 = -1$.

The second system $x_1 = 1, x_2 = 1, x_3 = 1, x_4 = 1$.

d) Part (c) requires more work.

8. (a) $a_{ij} + b_{ij} = b_{ij} + a_{ij}$ for $1 \leq i \leq n, \quad 1 \leq j \leq m$.

(b) $(a_{ij} + b_{ij}) + c_{ij} = a_{ij} + (b_{ij} + c_{ij})$ for $1 \leq i \leq n, \quad 1 \leq j \leq m$.

(c) $a_{ij} + 0 = 0 + a_{ij} = a_{ij}$ for $1 \leq i \leq n, \quad 1 \leq j \leq m$.

(d) $a_{ij} + (-a_{ij}) = -a_{ij} + a_{ij} = 0$ for $1 \leq i \leq n, \quad 1 \leq j \leq m$.

(e) $\lambda(a_{ij} + b_{ij}) = \lambda a_{ij} + \lambda b_{ij}$ for $1 \leq i \leq n, \quad 1 \leq j \leq m$.

(f) $(\lambda + \mu)a_{ij} = \lambda a_{ij} + \mu a_{ij}$ for $1 \leq i \leq n, \quad 1 \leq j \leq m$.

(g) $\lambda(\mu a_{ij}) = (\lambda\mu)a_{ij}$ for $1 \leq i \leq n, \quad 1 \leq j \leq m$.

(h) $1 a_{ij} = a_{ij}$ for $1 \leq i \leq n, \quad 1 \leq j \leq m$.

10. d) To find the inverse of the $n \times n$ matrix A:

INPUT $n \times n$ matrix $A = (a_{ij})$.

OUTPUT $n \times n$ matrix $B = A^{-1}$.

STEP 1. Initialize the $n \times n$ matrix $B = (b_{ij})$ to

$$b_{ij} = \begin{cases} 0 & i \neq j, \\ 1 & i = j. \end{cases}$$

STEP 2. For $i = 1, \ldots, n-1$ do STEPS 3, 4, and 5.

STEP 3. Let p be the smallest integer with $i \leq p \leq n$ and $a_{p,i} \neq 0$.

   If no integer p can be found then

     OUTPUT ('A is singular');

    STOP.

STEP 4. If $p \neq i$ then perform $(E_p) \leftrightarrow (E_i)$.

STEP 5. For $j = i+1, \ldots, n$ do STEPS 6 through 9.

  STEP 6. Set $m_{ji} = a_{ji}/a_{ii}$.

  STEP 7. For $k = i+1, \ldots, n$

    set $a_{jk} = a_{jk} - m_{ji}a_{ik}$; $a_{ij} = 0$.

  STEP 8. For $k = 1, \ldots, i-1$

    set $b_{jk} = b_{jk} - m_{ji}b_{ik}$.

  STEP 9. Set $b_{ji} = -m_{ji}$.

STEP 10. If $a_{nn} = 0$ then OUTPUT ('A is singular');

    STOP.

STEP 11. For $j = 1, \ldots, n$ do STEPS 12, 13 and 14.

  STEP 12. Set $b_{nj} = b_{nj}/a_{nn}$.

  STEP 13. For $i = n-1, \ldots, j$

    set $b_{ij} = (b_{ij} - \sum_{k=i+1}^{n} a_{ik}b_{kj})/a_{ii}$.

  STEP 14. For $i = j-1, \ldots, 1$

    set $b_{ij} = -[\sum_{k=i+1}^{n} a_{ik}b_{kj}]/a_{ii}$.

STEP 15. OUTPUT (B);

   STOP.

12. b) No, since the products $A_{ij}B_{jk}$ for $1 \leq i, j, k \leq 2$ cannot be formed.

 c) The following are necessary and sufficient conditions:

  i) The number of columns of A is the same as the number of rows of B.

  ii) The number of vertical lines of A equals the number of horizontal lines of B.

  iii) The placement of the vertical lines of A is identical to placement of the horizontal lines of B.

13. a) $a_{i1}, a_{i2}, \ldots, a_{im}$ represents the total number of plants of type v_i eaten by herbivores in the species $h_1, \ldots, h_m$ respectively. The number of herbivores of types $h_1, \ldots, h_m$ eaten by species c_j is $b_{1j}, \ldots, b_{mj}$, respectively. Thus, the total number of plants of type v_i ending up in species c_j is $a_{i1}b_{1j} + a_{i2}b_{2j} + \ldots + a_{im}b_{mj} = (AB)_{ij}$.

b) We first assume $n = m = k$ so that the matrices will have inverses. Let $x_1, \ldots, x_n$ represent the vegetations of type $v_1, \ldots, v_n$, let $y_1, \ldots, y_n$ represent the number of herbivores of species $h_1, \ldots, h_n$, and let $z_1, \ldots, z_n$ represent the number of carnivores of species $c_1, \ldots, c_n$.

If

$$
\begin{bmatrix} x_1 \\ x_2 \\ \vdots \\ x_n \end{bmatrix} = A \begin{bmatrix} y_1 \\ y_2 \\ \vdots \\ y_n \end{bmatrix},
$$

then

$$
\begin{bmatrix} y_1 \\ y_2 \\ \vdots \\ y_n \end{bmatrix} = A^{-1} \begin{bmatrix} x_1 \\ x_2 \\ \vdots \\ x_n \end{bmatrix}.
$$

Thus, $(A^{-1})_{i,j}$ represents the amount of type v_j plants eaten by a herbivore of species h_i. Similiarly, if

$$
\begin{bmatrix} y_1 \\ y_2 \\ \vdots \\ y_n \end{bmatrix} = B \begin{bmatrix} z_1 \\ z_2 \\ \vdots \\ z_n \end{bmatrix},
$$

then

$$
\begin{bmatrix} z_1 \\ z_2 \\ \vdots \\ z_n \end{bmatrix} = B^{-1} \begin{bmatrix} y_1 \\ y_2 \\ \vdots \\ y_n \end{bmatrix}.
$$

Thus, $(B^{-1})_{i,j}$ represents the number of herbivores of species h_j eaten by a carnivore of species c_i. If $x = Ay$ and $y = Bz$, then $x = ABz$ and $z = (AB)^{-1}x$. But, $y = A^{-1}x$ and $z = B^{-1}y$, so $z = B^{-1}A^{-1}x$.

Exercise Set 6.5 (PAGE 352)

2. The same answer as in Exercise 1.

4. Let

$$
A = \begin{bmatrix} a_{11} & a_{12} & a_{13} \\ a_{21} & a_{22} & a_{23} \\ a_{31} & a_{32} & a_{33} \end{bmatrix} \quad \text{and} \quad \tilde{A} = \begin{bmatrix} a_{21} & a_{22} & a_{23} \\ a_{11} & a_{12} & a_{13} \\ a_{31} & a_{32} & a_{33} \end{bmatrix}.
$$

Expanding along the third rows gives

$$
\det A = a_{31} \det \begin{bmatrix} a_{12} & a_{13} \\ a_{22} & a_{23} \end{bmatrix} - a_{32} \det \begin{bmatrix} a_{11} & a_{13} \\ a_{21} & a_{23} \end{bmatrix} + a_{33} \det \begin{bmatrix} a_{11} & a_{12} \\ a_{21} & a_{22} \end{bmatrix}
$$
$$
= a_{31}(a_{12}a_{23} - a_{13}a_{22}) - a_{32}(a_{11}a_{23} - a_{13}a_{21}) + a_{33}(a_{11}a_{22} - a_{12}a_{21})
$$

and

$$\det \tilde{A} = a_{31} \det \begin{bmatrix} a_{22} & a_{23} \\ a_{12} & a_{13} \end{bmatrix} - a_{32} \det \begin{bmatrix} a_{21} & a_{23} \\ a_{11} & a_{13} \end{bmatrix} + a_{33} \det \begin{bmatrix} a_{21} & a_{22} \\ a_{11} & a_{12} \end{bmatrix}$$

$$= a_{31}(a_{13}a_{22} - a_{12}a_{23}) - a_{32}(a_{13}a_{21} - a_{11}a_{23}) + a_{33}(a_{12}a_{21} - a_{11}a_{22}) = -\det A.$$

The other two cases are similiar.

5. When $n = 2$, $\det A = a_{11}a_{22} - a_{12}a_{21}$ requires 2 multiplications and 1 subtraction. Since

$$2! \sum_{k=1}^{1} \frac{1}{k!} = 2 \quad \text{and} \quad 2! - 1 = 1,$$

the formula holds for $n = 2$.

Assume the formula is true for $n = 2, ..., m$ and let A be an $(m+1) \times (m+1)$ matrix. Then

$$\det A = \sum_{j=1}^{m+1} a_{ij} A_{ij}$$

for any i, where $1 \leq i \leq m+1$.

To compute each A_{ij} requires

$$m! \sum_{k=1}^{m-1} \frac{1}{k!} \quad \text{multiplications} \quad \text{and} \quad m! - 1 \quad \text{additions/subtractions}.$$

Thus, the number of multiplications for $\det A$ is

$$(m+1)[m! \sum_{k=1}^{m-1} \frac{1}{k!}] + (m+1) = (m+1)! \left[\sum_{k=1}^{m-1} \frac{1}{k!} + \frac{1}{m!} \right] = (m+1)! \sum_{k=1}^{m} \frac{1}{k!}$$

and the number of additions/subtractions is

$$(m+1)[m! - 1] + m = (m+1)! - 1.$$

By the principle of mathematical induction the formula is valid for any $n \geq 2$.

6. The result follows from $\det AB = \det A \ \det B$ and Theorem 6.14.

8. a) If D_i is the determinant of the matrix formed by replacing the ith column of A with **b** and if $D = \det A$, then

$$x_i = D_i/D \quad \text{for } i = 1, \dots, n.$$

b) $(n+1)! \left(\sum_{k=1}^{n-1} \frac{1}{k!} \right) + n$ multiplications/divisions

$(n+1)! - n - 1$ additions/subtractions.

71

Exercise Set 6.6 (PAGE 363)

2. a) $x_1 = 1$, $x_2 = 2$, $x_3 = -1$

 b) $x_1 = 1.5$, $x_2 = 2$, $x_3 = -1.199998$, $x_4 = 3$

 c) $x_1 = 1$, $x_2 = 1$, $x_3 = 1$

 d) $x_1 = -3.44744$, $x_2 = 5.57458$, $x_3 = 3.21845$

 e) $x_1 = -0.5$, $x_2 = 1.16666$, $x_3 = 1.25$

 f) $x_1 = 2.939851$, $x_2 = 0.07067770$, $x_3 = 5.677735$, $x_4 = 4.379812$

4. Partition $A^{(n)}$ into the form

$$
A^{(k)} = \begin{bmatrix}
a_{11}^{(1)} & a_{12}^{(1)} & \cdots & a_{1,k}^{(1)} & a_{1,k+1}^{(1)} & \cdots & a_{1,n}^{(1)} \\
0 & a_{22}^{(2)} & \cdots & a_{2k}^{(2)} & a_{2,k+1}^{(2)} & \cdots & a_{2,n}^{(2)} \\
\vdots & \ddots & \ddots & \vdots & \vdots & \ddots & \vdots \\
0 & \cdots & 0 & a_{k,k}^{(k)} & a_{k,k+1}^{(k)} & \cdots & a_{k,n}^{(k)} \\
0 & \cdots & 0 & a_{k+1,k}^{(k)} & a_{k+1,k+1}^{(k)} & \cdots & a_{k+1,n}^{(k)} \\
\vdots & \ddots & \vdots & \vdots & \vdots & \ddots & \vdots \\
0 & \cdots & 0 & a_{n,k}^{(k)} & a_{n,k+1}^{(k)} & \cdots & a_{n,n}^{(k)}
\end{bmatrix}
= \begin{bmatrix} A_{11}^{(k)} & A_{12}^{(k)} \\ A_{21}^{(k)} & A_{22}^{(k)} \end{bmatrix}
$$

The multiplier matrix $M^{(k-1)}$ and $A^{(k-1)}$ can be similarly partitioned into

$$
M^{(k-1)} = \begin{bmatrix}
1 & 0 & \cdots & & \cdots & 0 & 0 & \cdots & \cdots & 0 \\
0 & \ddots & \ddots & & & \vdots & \vdots & & & \vdots \\
\vdots & \ddots & \ddots & & \ddots & \vdots & \vdots & & & \vdots \\
0 & \cdots & 0 & & 1 & 0 & \vdots & & & \vdots \\
0 & \cdots & 0 & & -m_{k,k-1} & 1 & 0 & \cdots & \cdots & 0 \\
0 & \cdots & 0 & & -m_{k+1,k-1} & 0 & 1 & 0 & \cdots & 0 \\
\vdots & & \vdots & & \vdots & \vdots & 0 & \ddots & \ddots & \vdots \\
\vdots & & \vdots & & \vdots & \vdots & \vdots & \ddots & \ddots & 0 \\
0 & \cdots & 0 & & -m_{n,k-1} & 0 & 0 & \cdots & 0 & 1
\end{bmatrix}
= \begin{bmatrix} M_{11}^{(k-1)} & O \\ M_{21}^{(k-1)} & I \end{bmatrix},
$$

where $M_{11}^{(k-1)}$ is a $k \times k$ lower triangular matrix, O is $k \times (n-k)$ block of zeros, $M_{21}^{(k-1)}$ is an $(n-k) \times k$ matrix, I is an $(n-k) \times (n-k)$ identity matrix, and

$$
A^{(k-1)} = \begin{bmatrix} A_{11}^{(k-1)} & A_{12}^{(k-1)} \\ A_{21}^{(k-1)} & A_{22}^{(k-1)} \end{bmatrix}.
$$

Here $A_{11}^{(k-1)}$ is $k \times k$, $A_{12}^{(k-1)}$ is $k \times (n-k)$, $A_{21}^{(k-1)}$ is $(n-k) \times k$ and $A_{22}^{(k-1)}$ is $(n-k) \times (n-k)$. The formation of $A_{11}^{(k)}$ can be obtained from the partitioned product of $M^{(k-1)}$ and $A^{(k-1)}$ and is given by

$$A_{11}^{(k)} = M_{11}^{(k-1)} A_{11}^{(k-1)} + 0.A_{21}^{(k-1)} = M_{11}^{(k-1)} A_{11}^{(k-1)}.$$

In a similiar manner, each of $M^{(k-2)}, \ldots, M^{(1)}$ and $A^{(k-2)}, \ldots, A^{(1)}$ can be partitioned to obtain

$$A_{11}^{(k)} = M_{11}^{(k-1)} A_{11}^{(k-1)} = M_{11}^{(k-1)} M_{11}^{(k-2)} A_{11}^{(k-2)} = \cdots = M_{11}^{(k-1)} M_{11}^{(k-2)} \ldots M_{11}^{(1)} A_{11}^{(1)},$$

where $A_{11}^{(1)} = A_{11}$ is the $k \times k$ leading principal submatrix of A.

Assume all leading principal submatrices of A are nonsingular. Then $a_{11} \neq 0$ and the elimination process can be started. For the inductive hypothesis, assume that $k-1$ elimination steps can be performed without row interchanges. It follows that $a_{11}^{(1)}, \ldots, a_{k-1,k-1}^{(k-1)}$ are all nonzero and the above equation holds. Taking determinants produces

$$a_{11}^{(1)} a_{22}^{(2)} \ldots a_{k-1,k-1}^{(k-1)} \cdot a_{k,k}^{(k)} = \det A_{11}^{(k)} = \det M_{11}^{(k-1)} \det M_{11}^{(k-2)} \ldots \det M_{11}^{(1)} \det A_{11} \neq 0.$$

Hence, $a_{k,k}^{(k)} \neq 0$ and the process can continue. By mathematical induction all pivot elements $a_{11}^{(1)}, \ldots, a_{n,n}^{(n)}$ are nonzero and Gaussian elimination can be performed without row interchanges.

Conversely, suppose Gaussian elimination can be performed without row interchanges. It follows that all the pivot elements $a_{11}^{(1)}, \ldots, a_{n,n}^{(n)}$ are nonzero. Thus,

$$\det A_{11} = a_{11}^{(1)} a_{22}^{(2)} \ldots a_{k,k}^{(k)} \neq 0,$$

and the $k \times k$ principal leading submatrix is nonsingular for each $k = 1, 2, \ldots, n$.

Exercise Set 6.7 (PAGE 377)

2. a) yes

b) Not necessarily. Consider

$$\begin{bmatrix} 2 & -1 \\ 3 & 4 \end{bmatrix}.$$

c) Not necessarily. Consider

$$\begin{bmatrix} 2 & 1 \\ 1 & 2 \end{bmatrix} \quad \text{and} \quad \begin{bmatrix} -2 & 1 \\ 1 & -2 \end{bmatrix}.$$

d) Not necessarily. Consider

$$\begin{bmatrix} 2 & -1 \\ 3 & 4 \end{bmatrix}.$$

e) Not necessarily. Consider

$$\begin{bmatrix} 2 & 1 \\ 1 & 2 \end{bmatrix} \quad \text{and} \quad \begin{bmatrix} 2 & -1 \\ -1 & 2 \end{bmatrix}.$$

4. a) $k = 2$

b) A cannot be strictly diagonally dominant regardless of k.

c) all values of k

d) $k > 2$

6. Yes

8. a) $\begin{bmatrix} 2 & -1 & -1 \\ -1 & 2 & -1 \\ -1 & -1 & 2 \end{bmatrix}$ b) No

10. a)

$$L = \begin{bmatrix} 1.41423 & 0 & 0 \\ -0.7071069 & 1.224743 & 0 \\ 0 & -0.8164972 & 1.154699 \end{bmatrix}$$

b)

$$L = \begin{bmatrix} 2 & 0 & 0 & 0 \\ 0.5 & 1.658311 & 0 & 0 \\ 0.5 & -0.7537785 & 1.087113 & 0 \\ 0.5 & 0.4522671 & 0.08362442 & 1.240346 \end{bmatrix}$$

c)

$$L = \begin{bmatrix} 2 & 0 & 0 & 0 \\ 0.5 & 1.658311 & 0 & 0 \\ -0.5 & -0.4522671 & 2.132006 & 0 \\ 0 & 0 & 0.9380833 & 1.766351 \end{bmatrix}$$

d)

$$L = \begin{bmatrix} 2.449489 & 0 & 0 & 0 \\ 0.8164966 & 1.825741 & 0 & 0 \\ 0.4082483 & 0.3651483 & 1.923538 & 0 \\ -0.4082483 & 0.1825741 & -0.4678876 & 1.606574 \end{bmatrix}$$

12. a) $1, -1, 0$

b) $0.2, -0.2, -0.2, 0.25$

c) $1, 2, -1, 2$

d) $-0.8586387, 2.418848, -0.9581152, -1.272251$

14. STEP 1. Set $l_1 = a_1; u_1 = c_1/l_1$.

STEP 2. For $i = 2, \ldots, n - 1$

74

set $l_i = a_i - b_i u_{i-1}; u_i = c_i/l_i$.

STEP 3. Set $l_n = a_n - b_n u_{n-1}$.

STEP 4. Set $z_1 = d_1/l_1$.

STEP 5. For $i = 2, \ldots, n$ set $z_i = (d_i - b_i z_{i-1})/l_i$.

STEP 6. Set $x_n = z_n$.

STEP 7. For $i = n - 1, \ldots, 1$ set $x_i = z_i - u_i x_{i+1}$.

STEP 8. OUTPUT $(x_1, \ldots, x_n)$;

      STOP.

16. a)

$$L = \begin{bmatrix} 1 & 0 & 0 \\ -1 & 1 & 0 \\ 2 & 1 & 1 \end{bmatrix}, \qquad D = \begin{bmatrix} 3 & 0 & 0 \\ 0 & -1 & 0 \\ 0 & 0 & 2 \end{bmatrix}$$

b)

$$L = \begin{bmatrix} 1 & 0 & 0 \\ -2 & 1 & 0 \\ 3 & -1 & 1 \end{bmatrix}, \qquad D = \begin{bmatrix} 3 & 0 & 0 \\ 0 & 1 & 0 \\ 0 & 0 & 0 \end{bmatrix}$$

c)

$$L = \begin{bmatrix} 1 & 0 & 0 & 0 \\ -2 & 1 & 0 & 0 \\ 0 & 2 & 1 & 0 \\ -1 & 1 & 4 & 1 \end{bmatrix}, \qquad D = \begin{bmatrix} -1 & 0 & 0 & 0 \\ 0 & 1 & 0 & 0 \\ 0 & 0 & 1 & 0 \\ 0 & 0 & 0 & -4 \end{bmatrix}$$

d)

$$L = \begin{bmatrix} 1 & 0 & 0 & 0 \\ -1 & 1 & 0 & 0 \\ 2 & 0 & 1 & 0 \\ -2 & 1 & -1 & 1 \end{bmatrix}, \qquad D = \begin{bmatrix} 2 & 0 & 0 & 0 \\ 0 & 1 & 0 & 0 \\ 0 & 0 & 2 & 0 \\ 0 & 0 & 0 & 3 \end{bmatrix}$$

18. First, $|l_{11}| = |a_{11}| > 0$ and $|u_{12}| = \frac{|a_{12}|}{|l_{11}|} < 1$.

In general, assume $|l_{jj}| > 0$ and $|u_{j,j+1}| < 1$ for $j = 1, \ldots, i-1$. Then

$$|l_{ii}| = |a_{ii} - l_{i,i-1} u_{i-1,i}| = |a_{ii} - a_{i,i-1} u_{i-1,i}| \geq |a_{ii}| - |a_{i,i-1} u_{i-1,i}| > |a_{ii}| - |a_{i,i-1}| > 0,$$

and

$$|u_{i,i+1}| = \frac{|a_{i,i+1}|}{|l_{ii}|} < \frac{|a_{i,i+1}|}{|a_{ii}| - |a_{i,i-1}|} \leq 1,$$

for $i = 2, \ldots, n - 1$.

Further,

$$|l_{nn}| = |a_{nn} - l_{n,n-1} u_{n-1,n}| = |a_{nn} - a_{n,n-1} u_{n-1,n}| \geq |a_{nn}| - |a_{n,n-1}| > 0.$$

So

$$\det A = \det L \det U = l_{11} \cdot l_{22} \ldots l_{nn} \cdot 1 > 0.$$

20. For this proof we need the following result, which can be found in *Applied Linear Algebra* by Noble and Daniel [93] (pages 210 and 452): If A is an $m \times m$ nonsingular matrix, D an $n \times n$ nonsingular matrix, B an $m \times n$ matrix, and C an $n \times m$ matrix, then

$$\det \begin{bmatrix} A & B \\ -C & D \end{bmatrix} = \det A \det(D + CA^{-1}B) = \det D \det(A + BD^{-1}C).$$

To establish the existence of the factorization, it suffices to show that L_i is nonsingular for $i = 1, \ldots, n$. Since $L_1 = A_1$ and A_1 is nonsingular, L_1 is nonsingular. Moreover, $L_2 = A_2 - B_2\Gamma_1 = A_2 - B_2 L_1^{-1}C_1$, so $\det L_2 = \det(A_2 - B_2 L_1^{-1}C_1)$. But,

$$0 \neq \det \begin{bmatrix} A_1 & C_1 \\ B_2 & A_2 \end{bmatrix} = \det A_1 \det(A_2 - B_2 A_1^{-1}C_1) = \det A_1 \det L_2$$

so L_2 is nonsingular.

Assume that $L_1, \ldots, L_k$ are nonsingular. Since

$$L_k = A_k - B_k\Gamma_{k-1} = A_k - B_k L_{k-1}^{-1}C_{k-1},$$

the following equation holds

$$\begin{bmatrix} A_1 & C_1 & & & \\ B_2 & A_2 & C_2 & & \\ & \ddots & \ddots & C_{k-1} \\ & & B_k & A_k \end{bmatrix} = \begin{bmatrix} L_1 & & & \\ B_2 & L_2 & & \\ & \ddots & \ddots & \\ & & B_k & L_k \end{bmatrix} \begin{bmatrix} I & L_1^{-1}C_1 & & \\ & I & L_2^{-1}C_2 & \\ & & \ddots & L_{k-1}^{-1}C_{k-1} \\ & & & I \end{bmatrix}.$$

Denote this equation by

$$A^{(k)} = L^{(k)}U^{(k)}.$$

Since $\det A^{(k)} \neq 0$ and $\det U^{(k)} = 1$, we have $\det L^{(k)} = \det A^{(k)} \neq 0$.
Now, $L_{k+1} = A_{k+1} - B_{k+1}\Gamma_k = A_{k+1} - B_{k+1}L_k^{-1}C_k$ can be formed.
Consider the equation

$$A^{(k+1)} = L^{(k+1)}U^{(k+1)},$$

or in the partitioned form

$$\begin{bmatrix} A^{(k)} & C \\ B & A_{k+1} \end{bmatrix} = \begin{bmatrix} L^{(k)} & 0 \\ B' & L_{k+1} \end{bmatrix} \begin{bmatrix} U^{(k)} & C' \\ 0 & I \end{bmatrix}.$$

By assumption, $\det A^{(k+1)} \neq 0$, so $\det L^{(k+1)} \neq 0$. But

$$\det L^{(k+1)} = \det L_{k+1} \det L^{(k)}$$

so L_{k+1} is nonsingular .

By induction it follows that L_k is nonsingular for $k = 1, \ldots, n$.

22. $5n - 4$ multiplications/divisions and $3n - 3$ subtractions/additions.

24. a) Mating male i with female j, or male j with female i, yields offspring with the same wing characteristics.

   b) No. Consider, for example, $\mathbf{x} = (1, 0, -1)^t$.

Chapter 7
Iterative Techniques in Matrix Algebra

Exercise Set 7.1 (PAGE 393)

2. a) Since $\|\mathbf{x}\|_1 = \sum_{i=1}^{n} |x_i| \geq 0$ with equality only if $x_i = 0$ for all i, properties (i) and (ii) in Definition 7.1 hold.

Also,

$$\|\alpha\mathbf{x}\|_1 = \sum_{i=1}^{n} |\alpha x_i| = \sum_{i=1}^{n} |\alpha||x_i| = |\alpha| \sum_{i=1}^{n} |x_i| = |\alpha|\|\mathbf{x}\|_1$$

so property (iii) holds.

Finally,

$$\|\mathbf{x} + \mathbf{y}\|_1 = \sum_{i=1}^{n} |x_i + y_i| \leq \sum_{i=1}^{n} (|x_i| + |y_i|) = \sum_{i=1}^{n} |x_i| + \sum_{i=1}^{n} |y_i| = \|\mathbf{x}\|_1 + \|\mathbf{y}\|_1,$$

so property (iv) also holds.

b) (1a) 8.5   (1b) 10   (1c) $|\sin k| + |\cos k| + e^k$ (1d) $4/(k+1) + 2/k^2 + k^2 e^{-k}$

4. Using Theorem 7.9 gives, for any positive integer, k

$$\|\mathbf{x}^{(k)} - \mathbf{x}\|_\infty \leq \|\mathbf{x}^{(k)} - \mathbf{x}\|_2 \leq \sqrt{n}\|\mathbf{x}^{(k)} - \mathbf{x}\|_\infty,$$

and the result follows easily.

6. a) $\|A\|_\infty = \max_{|x_1| \leq 1, |x_2| \leq 1}(|x_1 - x_2|, |2x_1 + x_2|) = \max(2, 3) = 3$
b) $\|A\|_\infty = \max_{|x_1| \leq 1, |x_2| \leq 1}(|x_1 + x_2|, |x_1 + x_2|) = 2$
c) $\|A\|_\infty = \max_{|x_1| \leq 1, |x_2| \leq 1}(|10x_1 + 15x_2|, |x_2|) = \max(25, 1) = 25$
d) $\|A\|_\infty = \max_{|x_1| \leq 1, |x_2| \leq 1}(|13x_1 + x_2|, |2x_1 + 5x_2|) = \max(4, 7) = 7$

7. Let $\mathbf{x}$ be a vector with $\|\mathbf{x}\|_1 = 1$. Then

$$\|A\mathbf{x}\|_1 = \sum_{i=1}^{n} |(A\mathbf{x})_i| = \sum_{i=1}^{n} \left| \sum_{j=1}^{n} a_{ij} x_j \right| \leq \sum_{i=1}^{n} \sum_{j=1}^{n} |a_{ij}||x_j|$$

$$= \sum_{j=1}^{n} \sum_{i=1}^{n} |a_{ij}||x_j| = \sum_{j=1}^{n} |x_j| \left(\sum_{i=1}^{n} |a_{ij}| \right)$$

$$\leq \sum_{j=1}^{n} |x_j| \left(\max_{1 \leq j \leq n} \sum_{i=1}^{n} |a_{ij}| \right)$$

$$= \|\mathbf{x}\|_1 \left(\max_{1 \leq j \leq n} \sum_{i=1}^{n} |a_{ij}| \right).$$

78

Thus,

$$\|A\mathbf{x}\|_1 \le \max_{1 \le j \le n} \sum_{i=1}^{n} |a_{ij}|,$$

and

$$\max_{\|\mathbf{x}\|_1 = 1} \|A\mathbf{x}\|_1 = \|A\|_1 \le \max_{1 \le j \le n} \sum_{i=1}^{n} |a_{ij}|.$$

Let p be an integer, $1 \le p \le n$, for which

$$\max_{1 \le j \le n} \sum_{i=1}^{n} |a_{ij}| = \sum_{i=1}^{n} |a_{ip}|.$$

Define $\mathbf{x} = (x_1, \ldots, x_n)^t$ by

$$x_i = \begin{cases} 0, & \text{if } i \ne p \\ 1, & \text{if } i \ne p. \end{cases}$$

Then $\|\mathbf{x}\|_1 = 1$ and $\|A\|_1 \ge \|A\mathbf{x}\|_1 = \sum_{i=1}^{n} |(A\mathbf{x})_i|$, so

$$\|A\|^{-1} = \sum_{i=1}^{n} |\sum_{j=1}^{n} a_{ij} x_j| = \sum_{i=1}^{n} |a_{ip} x_p| = \max_{1 \le j \le n} \sum_{i=1}^{n} |a_{ij}|.$$

8. (5a) 3   (5b) 2   (5c) 16   (5d) 6

9. a) Showing properties $(i) - (iv)$ of Definition 7.10 is similar to the proof in Excercise 2(a).

Property (v) is shown as follows:

$$\|AB\|_1 = \sum_{i=1}^{n} \sum_{j=1}^{n} |\sum_{k=1}^{n} a_{ik} b_{kj}| \le \sum_{i=1}^{n} \sum_{j=1}^{n} \sum_{k=1}^{n} |a_{ik}||b_{kj}|$$

$$= \sum_{i=1}^{n} \{ \sum_{k=1}^{n} (|a_{ik}| \sum_{j=1}^{n} |b_{kj}|) \}$$

$$\le \sum_{i=1}^{n} (\sum_{k=1}^{n} |a_{ik}|)(\sum_{k=1}^{n} \sum_{j=1}^{n} |b_{kj}|)$$

$$= \left(\sum_{i=1}^{n} \sum_{k=1}^{n} |a_{ik}| \right) \|B\|_1 = \|A\|_1 \|B\|_1.$$

10. a) Showing properties $(i) - (iv)$ of Definition 7.10 is similar to the proof of Theorem 7.4.

Property (v) is shown as follows:

$$\|AB\|_F^2 = (\sum_{i=1}^{n}\sum_{j=1}^{n}|\sum_{k=1}^{n}a_{ik}b_{kj}|^2)$$

$$\leq (\sum_{i=1}^{n}\sum_{j=1}^{n}(\sum_{k=1}^{n}|a_{ik}|^2\sum_{k=1}^{n}|b_{kj}|^2)) \quad \text{by Lemma 8.5}$$

$$= \sum_{i=1}^{n}\sum_{k=1}^{n}\left[|a_{ik}|^2(\sum_{j=1}^{n}\sum_{k=1}^{n}|b_{kj}|^2)\right]$$

$$= \sum_{i=1}^{n}\sum_{k=1}^{n}|a_{ik}|^2\|B\|_F^2 = \|B\|_F^2\sum_{i=1}^{n}\sum_{k=1}^{n}|a_{ik}|^2$$

$$= \|B\|_F^2\|A\|_F^2 = \|A\|_F^2\|B\|_F^2.$$

b) (5a) $\sqrt{7}$ (5b) 2 (5c) $\sqrt{326}$ (5d) $\sqrt{39}$

c)

$$\|A\|_2^2 = \max_{\|\mathbf{x}\|_2=1}\sum_{i=1}^{n}(\sum_{j=1}^{n}a_{ij}x_j)^2$$

$$\leq \max_{\|\mathbf{x}\|_2=1}\sum_{i=1}^{n}(\sum_{j=1}^{n}|a_{ij}||x_j|)^2$$

$$\leq \max_{\|\mathbf{x}\|_2=1}\sum_{i=1}^{n}[(\sum_{i=1}^{n}|a_{ij}|^2)^{\frac{1}{2}}(\sum_{j=1}^{n}|x_j|^2)^{\frac{1}{2}}]^2$$

$$= \max_{\|\mathbf{x}\|_2=1}\sum_{i=1}^{n}(\sum_{j=1}^{n}|a_{ij}|^2)(\sum_{j=1}^{n}|x_j|^2)$$

$$= \sum_{i=1}^{n}\sum_{j=1}^{n}|a_{ij}|^2$$

$$= \|A\|_F^2$$

Let j be fixed and define

$$x_k = \begin{cases} 0, & \text{if } k \neq j \\ 1, & \text{if } k = j. \end{cases}$$

Then $A\mathbf{x} = (a_{1j}, a_{2j}, \dots, a_{nj})^t$, so

$$\|A\|_2^2 \geq \|A\mathbf{x}\|_2^2 \geq \sum_{i=1}^{n}|a_{ij}|^2.$$

Thus,

$$\|A\|_F^2 = \sum_{i=1}^{n}\sum_{j=1}^{n}|a_{ij}|^2 = \sum_{j=1}^{n}\sum_{i=1}^{n}|a_{ij}|^2 \leq \sum_{j=1}^{n}\|A\|_2^2 = n\|A\|_2^2.$$

Hence, $\|A\|_2 \leq \|A\|_F \leq \sqrt{n}\|A\|_2$.

12. Clearly i) holds. If $||A|| = 0$, then $||A\mathbf{x}|| = 0$ for all vectors $\mathbf{x}$ with $||\mathbf{x}|| = 1$. Using $\mathbf{x} = (1, 0, \ldots, 0)^t$, $\mathbf{x} = (0, 1, 0, \ldots, 0)^t, \ldots$, and $\mathbf{x} = (0, \ldots, 0, 1)^t$ successively implies that each column of A is zero. Thus, $||A|| = 0$ if and only if $A = 0$.

Moreover,

$$||\alpha A|| = \max_{||\mathbf{x}||=1} ||(\alpha A\mathbf{x})|| = |\alpha| \max_{||\mathbf{x}||=1} ||A\mathbf{x}|| = |\alpha| \cdot ||A||,$$

$$||A + B|| = \max_{||\mathbf{x}||=1} ||(A + B)\mathbf{x}|| \leq \max_{||\mathbf{x}||=1} (||A\mathbf{x}|| + ||B\mathbf{x}||)$$

$$\leq \max_{||\mathbf{x}||=1} ||A\mathbf{x}|| + \max_{||\mathbf{x}||=1} ||B\mathbf{x}|| = ||A|| + ||B||,$$

and

$$||AB|| = \max_{||\mathbf{x}||=1} ||(AB)\mathbf{x}|| = \max_{||\mathbf{x}||=1} ||A(B\mathbf{x})||$$

$$\leq \max_{||\mathbf{x}||=1} ||A|| \, ||B\mathbf{x}|| = ||A|| \max_{||\mathbf{x}||=1} ||B\mathbf{x}|| = ||A|| \, ||B||.$$

14. Since $||\mathbf{x}||' = 0$ implies $||S\mathbf{x}|| = 0$, we have $S\mathbf{x} = 0$. Since S is nonsingular, $\mathbf{x} = 0$. Also,

$$||\mathbf{x} + \mathbf{y}||' = ||S(\mathbf{x} + \mathbf{y})|| = ||S\mathbf{x} + S\mathbf{y}|| \leq ||S\mathbf{x}|| + ||S\mathbf{y}|| = ||\mathbf{x}||' + ||\mathbf{y}||'$$

and

$$||\alpha\mathbf{x}||' = ||S(\alpha\mathbf{x})|| = |\alpha| \, ||S\mathbf{x}|| = |\alpha| \, ||\mathbf{x}||'.$$

Exercise Set 7.2 (PAGE 398)

2. a) 3 b) 1 c) 0.5 d) 5 e) 3 f) 7

4. Since

$$A_1^k = \begin{bmatrix} 1 & 0 \\ \frac{2^k-1}{2^{k+1}} & 2^{-k} \end{bmatrix}, \text{ we have } \lim_{k \to \infty} A_1^k = \begin{bmatrix} 1 & 0 \\ \frac{1}{2} & 0 \end{bmatrix}.$$

Also

$$A_2^k = \begin{bmatrix} 2^{-k} & 0 \\ \frac{16k}{2^{k-1}} & 2^{-k} \end{bmatrix}, \text{ so } \lim_{k \to \infty} A_2^k = \begin{bmatrix} 0 & 0 \\ 0 & 0 \end{bmatrix}.$$

6. a) Let $\mathbf{x} \neq 0$ be given and let $\mathbf{y} = \mathbf{x}/||\mathbf{x}||$. Then $||\mathbf{y}|| = 1$, so by Theorem 7.11,

$$||A|| \geq ||A\mathbf{x}||/||\mathbf{x}||.$$

b) If $\mathbf{y}$ is an eigenvector, then $\mathbf{x} = \frac{\mathbf{y}}{||\mathbf{y}||}$ is also an eigenvector.

7. See Noble and Daniel [93], pages 264 and 265, for a solution to this exercise.

8. a) $P(\lambda) = (\lambda_1 - \lambda)\ldots(\lambda_n - \lambda) = \det(A - \lambda I)$, so $P(0) = \lambda_1 \cdots \lambda_n = \det A$.

b) A singular if and only if $\det A = 0$, which is equivalent to at least one of λ_i being 0.

10. a) $\det(A - \lambda I) = \det((A - \lambda I)^t) = \det(A^t - \lambda I)$.

b) If $A\mathbf{x} = \lambda\mathbf{x}$, then $A^2\mathbf{x} = \lambda A\mathbf{x} = \lambda^2\mathbf{x}$ and by induction $A^k\mathbf{x} = \lambda^k\mathbf{x}$.

c) If $A\mathbf{x} = \lambda\mathbf{x}$ and A^{-1} exists, then $\mathbf{x} = \lambda A^{-1}\mathbf{x}$. By Exercise 8 b), $\lambda \neq 0$, so $\frac{1}{\lambda}\mathbf{x} = A^{-1}\mathbf{x}$.

d) Since $A^{-1}\mathbf{x} = \frac{1}{\lambda}\mathbf{x}$, $(A^{-1})^2\mathbf{x} = \frac{1}{\lambda}A^{-1}\mathbf{x} = \frac{1}{\lambda^2}\mathbf{x}$. Mathematical induction gives

$$(A^{-1})^k\mathbf{x} = \frac{1}{\lambda^k}\mathbf{x}.$$

e) If $A\mathbf{x} = \lambda\mathbf{x}$, then

$$q(A)\mathbf{x} = q_0\mathbf{x} + q_1 A\mathbf{x} + \ldots + q_k A^k\mathbf{x} = q_0\mathbf{x} + q_1\lambda\mathbf{x} + \ldots + q_k\lambda^k\mathbf{x} = q(\lambda)\mathbf{x}.$$

f) Let $A - \alpha I$ be nonsingular. Since $A\mathbf{x} = \lambda\mathbf{x}$,

$$(A - \alpha I)\mathbf{x} = A\mathbf{x} - \alpha I\mathbf{x} = \lambda\mathbf{x} - \alpha\mathbf{x} = (\lambda - \alpha)\mathbf{x}.$$

Thus,

$$\frac{1}{\lambda - \alpha}\mathbf{x} = (A - \alpha I)^{-1}\mathbf{x}.$$

12. Let $A\mathbf{x} = \lambda\mathbf{x}$. Then $|\lambda|\ \|\mathbf{x}\| = \|A\mathbf{x}\| \leq \|A\|\ \|\mathbf{x}\|$ which implies $|\lambda| \leq \|A\|$. Also, $\frac{1}{\lambda}\mathbf{x} = A^{-1}\mathbf{x}$ so $1/|\lambda| \leq \|A^{-1}\|$ and $\|A^{-1}\|^{-1} \leq |\lambda|$.

Exercise Set 7.3 (PAGE 413)

2. a) Divergence.
b) The Jacobi method cannot be used.
c) $\mathbf{x}^{(5)} = (0.9955500, 0.9572499, 0.7910999)^t$
d) $\mathbf{x}^{(5)} = (1.5, 2.0, -1.2, 3)^t$
e) Divergence.
f) $\mathbf{x}^{(7)} = (1.10322, 2.996505, -1.02175, -2.62411)^t$
g) $\mathbf{x}^{(11)} = (0.492284, 0.990045, 1.98245, -1.98232)^t$
h) $\mathbf{x}^{(12)} = (0.998006, 1.99436, 0.998006, 1.99601, 0.997179, 1.99601)^t$

4. a) Divergence.

b) The Gauss-seidel method cannot be used.

c) $\mathbf{x}^{(4)} = (0.9957476, 0.9578738, 0.7915747)^t$

d) $\mathbf{x}^{(2)} = (1.5, 2, -1.2, 3)^t$

e) Divergence.

f) $\mathbf{x}^{(4)} = (1.10290, 2.99569, -1.02101, -2.62008)^t$

g) $\mathbf{x}^{(7)} = (0.493283, 0.994251, 1.99346, -1.99564)^t$

h) $\mathbf{x}^{(7)} = (0.996554, 1.99706, 0.998745, 1.99792, 0.998225, 1.99924)^t$

6. a) $\omega = 0.5$; $\mathbf{x}^{(9)} = (0.990010, 0.948966, 0.785492)^t$
$\omega = 1.1$; $\mathbf{x}^{(4)} = (0.995253, 0.957979, 0.791627)^t$

b) $\omega = 1.012822$; $\mathbf{x}^{(4)} = (0, 9957846, 0.9578934, 0.7915787,)^t$

c) $\omega = 1.18691$; $\mathbf{x}^{(5)} = (0.5034478, 1.001558, 2.001185, -2.000466)^t$

d) $\omega = 1.334$; $\mathbf{x}^{(7)} = (1.00127, 2.00101, 1.00121, 1.99979, 0.999873, 1.99967)^t$
$\omega = 1.95$; divergence
$\omega = 0.95$; $\mathbf{x}^{(8)} = (0.996689, 1.99695, 0.998599, 1.99785, 0.998019, 1.99909)^t$

7. Subtract $\mathbf{x} = T\mathbf{x} + \mathbf{c}$ from $\mathbf{x}^{(k)} = T\mathbf{x}^{(k-1)} + \mathbf{c}$ to obtain $\mathbf{x}^{(k)} - \mathbf{x} = T(\mathbf{x}^{(k-1)} - \mathbf{x})$.
Thus

$$\|\mathbf{x}^{(k)} - \mathbf{x}\| \le \|T\| \, \|\mathbf{x}^{(k-1)} - \mathbf{x}\|.$$

Inductively, we have

$$\|\mathbf{x}^{(k)} - \mathbf{x}\| \le \|T\|^k \|\mathbf{x}^{(0)} - \mathbf{x}\|.$$

The remainder of the proof is similar to the proof of Corollary 2.5.

8. $T_j = (t_{ij})$ has entries given by

$$t_{ij} = \begin{cases} 0, & i = k \text{ for } 1 \le i, k \le n \\ \frac{-a_{ik}}{a_{ii}}, & i \ne k \text{ for } 1 \le i, k \le n. \end{cases}$$

Thus,

$$\|T_j\|_\infty = \max_{1 \le i \le n} \sum_{\substack{k=1 \\ k \ne 1}}^{n} |\frac{a_{ik}}{a_{ii}}| < 1,$$

since A is strictly diagonally dominant.

9. Let $\lambda_1, ..., \lambda_n$ be the eigenvalues of T_ω. Then

$$\prod_{i=1}^{n} \lambda_i = \det T_\omega$$

$$= \det \left((D - \omega L)^{-1}[(1 - \omega)D + \omega U] \right)$$

$$= \det(D - \omega L)^{-1} \det((1 - \omega)D + \omega U)$$

$$= \det(D^{-1}) \det((1 - \omega)D)$$

$$= \left(\frac{1}{(a_{11}a_{22} \dots a_{nn})} \right) \left((1 - \omega)^n a_{11}a_{22} \dots a_{nn}) \right)$$

$$= (1 - \omega)^n.$$

Thus,

$$\rho(T_\omega) = \max_{1 \le i \le n} |\lambda_i| \ge |\omega - 1|,$$

and $|\omega - 1| < 1$ if and only if $0 < \omega < 2$.

10. The iteration matrix

$$(D - L)^{-1}U = \begin{bmatrix} 0 & -1 & 0 & 0 & 0 \\ 0 & -1 & -1 & 0 & 0 \\ 0 & -0.7142857 & -0.7142857 & -0.2857143 & 0 \\ 0 & -0.7142857 & -0.7142857 & -0.2857143 & 1 \\ 0 & -0.4761905 & -0.4761905 & -0.1904762 & -0.\overline{6} \end{bmatrix}$$

has $\rho((D - L)^{-1}U) > 1$ so neither the Jacobi method nor the Gauss-Seidel method can be used.

Exercise Set 7.4 (PAGE 422)

2. a) 60,002
   c) 235.23
   e) 12

   b) 241.37
   d) 339,866
   f) 32

4. a) With $B = \begin{bmatrix} 1 & 2 \\ 1 & 2 \end{bmatrix}$, we have $K_\infty(A) \ge 30,001$.

   b) With $B = \begin{bmatrix} 4.0 & 1.6 \\ 7.0 & 2.8 \end{bmatrix}$, we have $K_\infty(A) \ge \frac{97}{3}$.

6. $(1.818192, 0.5909091)^t$; A is ill-conditioned since A is nearly singular.

8. $1.01(n^3 + 3n^2)$ is closer.

Chapter 8
Approximation Theory

Exercise Set 8.1 (PAGE 435)

2. $P_3(x) = 0.9999071 + 1.014109x + 0.4252571x^2 + 0.27893333x^3$ has error 6×10^{-7}.

$P_4(x) = 1.000000 + 0.9986333x + 0.5100667x^2 - 0.1402667x^3 + 0.06933333x^4$ has error 0.

4. $P_1(x) = 0.9295140 + 0.5281021x$ has error 2.457×10^{-2}.

$P_2(x) = 1.011341 - 0.3256988x + 1.147330x^2$ has error 9.453×10^{-4}.

$P_3(x) = 1.000440 - 0.001540986x - 0.011505675x^2 + 1.021023x^3$ has error 1.112×10^{-4}.

$P_4(x) = 0.9994951 + 0.1106990x - 0.8254245x^2 + 2.782884x^3 - 1.167916x^4$ has error 8.608×10^{-5}.

6. a) $1.665540x - 0.5124568$ has error 0.33559.

b) $1.129424x^2 - 0.3114035x + 0.08514401$ has error 2.4199×10^{-3}.

c) $0.2662081x^3 + 0.4029322x^2 + 0.2483857x - 0.01840140$ has error 5.0747×10^{-6}.

d) $0.04570748e^{2.707295x}$ has error 1.0750.

e) $0.9501565x^{1.872009}$ has error 0.054477.

8. $0.22335x - 0.80283$.

For minimal A, 406; for minimal D, 272. The prediction for A is not reasonable.

10. $0.17952x + 8.2084$

12. $1.600393x + 25.92175$

14. For each $i = 1, ..., n+1$ and $j = 1, ..., n+1$, $a_{ij} = a_{ji} = \sum_{k=1}^{m} x_k^{i+j-2}$, so $A = (a_{ij})$ is symmetric.

Suppose A is singular and $\mathbf{c} \neq 0$ satisfies $\mathbf{c}^t A \mathbf{c} = 0$. Then,

$$0 = \sum_{i=1}^{n+1}\sum_{j=1}^{n+1} a_{ij} c_i c_j = \sum_{i=1}^{n+1}\sum_{j=1}^{n+1} (\sum_{k=1}^{m} x_k^{i+j-2}) c_i c_j = \sum_{k=1}^{m} \left[\sum_{i=1}^{n+1}\sum_{j=1}^{n+1} c_i c_j x_k^{i+j-2} \right],$$

so

$$\sum_{k=1}^{m} (\sum_{i=1}^{n+1} c_i x_k^{i-1})^2 = 0.$$

Define $P(x) = c_1 + c_2 x + ... + c_{n+1} x^n$. Then $\sum_{k=1}^{m} [P(x_k)]^2 = 0$ and P has roots $x_1, ..., x_m$. Since the roots are distinct and $m > n$, P must be the zero polynomial. Thus, $c_1 = c_2 = ... = c_{n+1} = 0$ and A must be nonsingular.

Exercise Set 8.2 (PAGE 449)

2. a) $x^2 - 2x + 3$

b) $3x^2 - 2.4x - 0.6$

c) $0.1588765x^2 - 0.9313425x + 1.723526$

d) $0.3087217x^2 - 0.9305517x + 0.9944892$

e) $-2.431706x + 1.215853$

f) $-0.2335094x^2 + 1.382762x - 1.142993$

4. a) $x^2 - 2x + 3$ has error 0.

b) $\frac{3}{5}x + 1$ has error $\frac{8}{175}$.

d) $0.5367215x^2 - 1.103638x + 0.9962940$ has error 0.0014408.

e) $-2.279727x^2 + 0.7599089$ has error 0.076089.

6. $S_n(x) = \sum_{k=1}^{n-1} \frac{2}{k\sqrt{\pi}} (1 - (-1)^k) \sin kx$.

8. $L_1(x) = x - 1, L_2(x) = x^2 - 4x + 2, L_3(x) = x^3 - 9x^2 + 18x - 6.$

10. The normal equations are given by

$$\sum_{k=0}^{n} a_k \int_a^b x^{j+k} dx = \int_a^b x^j f(x) dx, j = 0, 1, \ldots, n.$$

Let

$$b_{jk} = \int_a^b x^{j+k} dx \quad \text{for each } j = 0, \ldots, n, \ k = 0, \ldots, n$$

and let $B = (b_{jk})$. Further, let

$$\mathbf{a} = (a_0, \ldots, a_n)^t \quad \text{and} \quad \mathbf{g} = \left(\int_a^b f(x) dx, \ldots, \int_a^b x^n f(x) dx \right)^t.$$

Then the normal equations produce the linear system $B\mathbf{a} = \mathbf{g}$. To show that the normal equations have a unique solution, it suffices to show that if $f \equiv 0$ then $\mathbf{a} = 0$. If $f \equiv 0$,

$$\sum_{k=0}^{n} a_k \int_a^b x^{j+k} dx = 0 \quad \text{for } j = 0, \ldots, n$$

and

$$\sum_{k=0}^{n} a_j a_k \int_a^b x^{j+k} dx = 0 \quad \text{for } j = 0, \ldots, n$$

and summing over j gives

$$\sum_{j=0}^{n} \sum_{k=0}^{n} a_j a_k \int_a^b x^{j+k} dx = 0.$$

86

Thus,

$$\int_a^b \sum_{j=0}^n \sum_{k=0}^n a_j x^j a_k x^k \, dx = 0$$

and

$$\int_a^b \left(\sum_{j=0}^n a_j x^j \right)^2 dx = 0.$$

Define $P(x) = a_0 + a_1 x + \cdots + a_n x^n$. Then $\int_a^b [P(x)]^2 dx = 0$ and $P(x) \equiv 0$. This implies that $a_0 = a_1 = \cdots = a_n = 0$, a contradiction. Hence, the matrix B is nonsingular and the normal equations have a unique solution.

11. Let $\phi_i(x) = \sum_{k=0}^n b_{i,k} x^k$ for each $i = 0, 1, \ldots, n$. Then $b_{ii} \neq 0$, since $\{\phi_0, \ldots, \phi_n\}$ is linearly independent.

Let $Q(x) = \sum_{k=0}^n a_k x^k$. Then $c_n = a_n / b_{n,n}$ and $c_i = (a_i - \sum_{j=i+1}^n c_j b_{j,i})/b_{i,i}$ for each $i = n-1, n-2, \ldots, 0$.

12. If $\sum_{i=0}^n c_i \phi_i(x) = 0$ for all $a \le x \le b$, then

$$\int_a^b \left(\sum_{i=0}^n c_i \phi_i(x) \right) \phi_j(x) w(x) \, dx = 0 \quad \text{for each } j = 0, 1, \ldots, n.$$

Thus, $c_j = 0$ for each $j = 0, 1, \ldots, n$.

13. The following integrations establish the orthogonality.

$$\int_{-\pi}^{\pi} [\phi_0(x)]^2 dx = \frac{1}{2\pi} \int_{-\pi}^{\pi} dx = 1,$$

$$\int_{-\pi}^{\pi} [\phi_k(x)]^2 dx = \frac{1}{\pi} \int_{-\pi}^{\pi} (\cos kx)^2 dx = \frac{1}{\pi} \int_{-\pi}^{\pi} \left[\frac{1}{2} + \frac{1}{2} \cos 2kx \right] dx = 1 + \left[\frac{1}{4k\pi} \sin 2kx \right]_{-\pi}^{\pi} = 1,$$

$$\int_{-\pi}^{\pi} [\phi_{n+k}(x)]^2 dx = \frac{1}{\pi} \int_{-\pi}^{\pi} (\sin kx)^2 dx = \frac{1}{\pi} \int_{-\pi}^{\pi} \left[\frac{1}{2} - \frac{1}{2} \sin 2kx \right] dx = 1 + \left[\frac{1}{4k\pi} \cos 2kx \right]_{-\pi}^{\pi} = 1,$$

$$\int_{-\pi}^{\pi} \phi_k(x)\phi_0(x) dx = \frac{1}{\sqrt{2\pi}} \int_{-\pi}^{\pi} \cos kx \, dx = \frac{1}{\sqrt{2k\pi}} \sin kx \Big]_{-\pi}^{\pi} = 0,$$

$$\int_{-\pi}^{\pi} \phi_{n+k}(x)\phi_0(x) dx = \frac{1}{\sqrt{2\pi}} \int_{-\pi}^{\pi} \sin kx \, dx = \frac{-1}{\sqrt{2k\pi}} \cos kx \Big|_{-\pi}^{\pi} = \frac{-1}{\sqrt{2k\pi}} [\cos k\pi - \cos(-k\pi)] = 0,$$

$$\int_{-\pi}^{\pi} \phi_k(x)\phi_j(x) dx = \frac{1}{\pi} \int_{-\pi}^{\pi} \cos kx \cos jx \, dx = \frac{1}{2\pi} \int_{-\pi}^{\pi} [\cos(k+j)x + \cos(k-j)x] dx = 0,$$

87

$$\int_{-\pi}^{\pi} \phi_{n+k}(x)\phi_{n+j}(x)dx = \frac{1}{\pi}\int_{-\pi}^{\pi} \sin kx \sin jx dx = \frac{1}{2\pi}\int_{-\pi}^{\pi}[\cos(k-j)x - \cos(k+j)x]dx = 0,$$

and

$$\int_{-\pi}^{\pi} \phi_k(x)\phi_{n+j}(x)dx = \frac{1}{\pi}\int_{-\pi}^{\pi} \cos kx \sin jx dx = \frac{1}{2\pi}\int_{-\pi}^{\pi}[\sin(k+j)x - \sin(k-j)x]dx = 0.$$

14.

$$\int_{-1}^{1} \frac{T_n^2(x)}{\sqrt{1-x^2}}\, dx = \int_{-1}^{1} \frac{\cos^2(n \arccos x)}{\sqrt{1-x^2}}\, dx$$

Perform the change of variable $x = \cos\theta$ to obtain

$$\int_{-1}^{1} \frac{T_n^2(x)}{\sqrt{1-x^2}}\, dx = \int_{0}^{\pi} \cos^2(n\theta)\, d\theta = \frac{\pi}{2}.$$

Exercise Set 8.3 (PAGE 459)

2. a) $x_0 = 0.9238795, x_1 = 0.3826834, x_2 = -0.3826834, x_3 = -0.9238795,$

$$P_3(x) = 2.519044 + 1.945377(x - x_0) + 0.7047420(x - x_0)(x - x_1)$$
$$+ 0.1751757(x - x_0)(x - x_1)(x - x_2).$$

b) $x_0 = 3.022023, x_1 = 2.171914, x_2 = 0.9696786, x_3 = 0.1195698,$

$$P_3(x) = 0.1192850 + 0.8297984(x - x_0) - 0.4043173(x - x_0)(x - x_1).$$

c) $x_0 = 1.961940, x_1 = 1.691342, x_2 = 1.308658, x_3 = 1.038060,$

$$P_3(x) = 0.6739337 + 0.5484575(x - x_0) - 0.1865365(x - x_0)(x - x_1)$$
$$+ 0.1058438(x - x_0)(x - x_1)(x - x_2).$$

d) $x_0 = 0.9619398, x_1 = 0.6913417, x_2 = 0.3086583, x_3 = 0.0380602,$

$$P_3(x) = 1.961938 + (x - x_0).$$

4. $x_0 = 0.9619398, x_1 = 0.6913417, x_2 = 0.3086583, x_3 = 0.0380602,$

$$P_3(x) = 1.400693 + 0.3702047(x - x_0) - 0.05951487(x - x_0)(x - x_1)$$
$$+ 0.02385053(x - x_0)(x - x_1)(x - x_2).$$

88

$P_3(0.1) = 1.048727$

6. $P(x) = \frac{383}{384} - \frac{4}{32}x^3$ has error at most 7.19×10^{-4}.

7. We have $T_0(x) = 1, T_1(x) = x$, and

$$T_{n+1}(x) = 2xT_{n-1}(x), \quad \text{for each } n = 1, 2, \ldots.$$

Thus,

$$T_2(x) = 2x^2 - 1;$$
$$T_3(x) = 4x^3 - 3x;$$
$$T_4(x) = 8x^4 - 8x^2 + 1;$$
$$T_5(x) = 16x^5 - 20x^3 + 5x;$$
$$T_6(x) = 32x^6 - 48x^4 + 18x^2 - 1;$$
$$T_7(x) = 64x^7 - 112x^5 + 56x^3 - 7x.$$

Solving for powers of x we have,

$$1 = T_0(x);$$

$$x = T_1(x);$$

$$x^2 = \frac{1}{2}T_2(x) + \frac{1}{2} = \frac{1}{2}T_2(x) + \frac{1}{2}T_0(x);$$

$$x^3 = \frac{1}{4}T_3(x) + \frac{3}{4}x = \frac{1}{4}T_3(x) + \frac{3}{4}T_1(x);$$

$$x^4 = \frac{1}{8}T_4(x) + x^2 - \frac{1}{8} = \frac{1}{8}T_4(x) + \frac{1}{2}T_2(x) + \frac{1}{2}T_0(x) - \frac{1}{8}T_0(x)$$
$$= \frac{1}{8}T_4(x) + \frac{1}{2}T_2(x) + \frac{3}{8}T_0(x);$$

$$x^5 = \frac{1}{16}T_5(x) + \frac{5}{4}x^3 - \frac{5}{16}x = \frac{1}{16}T_5(x) + \frac{5}{16}T_3(x) + \frac{15}{16}T_1(x) - \frac{5}{16}T_1(x)$$
$$= \frac{1}{16}T_5(x) + \frac{5}{16}T_3(x) + \frac{5}{8}T_1(x);$$

$$x^6 = \frac{1}{32}T_6(x) + \frac{3}{2}x^4 - \frac{9}{16}x^2 + \frac{1}{32}$$
$$= \frac{1}{32}T_6(x) + \frac{3}{16}T_4(x) + \frac{3}{4}T_2(x) + \frac{9}{16}T_0(x) - \frac{9}{32}T_2(x) - \frac{9}{32}T_0(x) + \frac{1}{32}T_0(x)$$
$$= \frac{1}{32}T_6(x) + \frac{3}{16}T_4(x) + \frac{15}{32}T_2(x) + \frac{5}{16}T_0(x);$$

$$x^7 = \frac{1}{64}T_7(x) + \frac{7}{4}x^5 - \frac{7}{8}x^3 + \frac{7}{64}x$$
$$= \frac{1}{64}T_7(x) + \frac{7}{64}T_5(x) + \frac{35}{64}T_3(x) + \frac{35}{32}T_1(x) - \frac{7}{32}T_3(x) - \frac{21}{32}T_1(x) + \frac{7}{64}T_1(x)$$
$$= \frac{1}{64}T_7(x) + \frac{7}{64}T_5(x) + \frac{21}{64}T_3(x) + \frac{35}{32}T_1(x).$$

8. If $i > j$, then

$$\frac{1}{2}(T_{i+j}(x) + T_{i-j}(x)) = \frac{1}{2}\left(\cos(i+j)\theta + \cos(i-j)\theta\right) = \cos i\theta \cos j\theta = T_i(x)T_j(x).$$

Exercise Set 8.4 (PAGE 465)

2.

$$r(x) = \frac{1 + \frac{3}{5}x + \frac{3}{20}x^2 + \frac{1}{60}x^3}{1 - \frac{2}{5}x + \frac{1}{20}x^2}$$

4. a)

$$r(x) = \frac{x}{1 + \frac{1}{6}x^2 + \frac{7}{360}x^4}$$

b)

$$r(x) = \frac{x - \frac{7}{60}x^3}{1 + \frac{1}{20}x^2}$$

6. a)

$$1 + \frac{4}{x - \frac{5}{4} + \frac{\frac{21}{16}}{x + \frac{1}{4}}}$$

b)

$$x - 2 + \cfrac{4}{x - 1.\overline{4} + \cfrac{2.25}{x + 3.1780688 + \cfrac{4.9475308}{x - 1.4740115 + \cfrac{2.0428833}{x - 0.25961288}}}}$$

c)

$$3x - \cfrac{15}{x - 0.4\overline{6} + \cfrac{6.15\overline{1}}{x + 0.38574181 - \cfrac{0.64446273}{x + 0.3759574 - \cfrac{0.20916703}{x - 0.29503254}}}}$$

d)

$$\cfrac{2}{3 + \cfrac{0.5}{x - 0.5 + \cfrac{7}{x + 1.2857142 - \cfrac{6.6326330}{x - 0.28571428}}}}$$

8. a) $0.8800998T_1 - 0.03912761T_3 + 0.0004995155T_5 - 0.000003004652T_7$

b) $1.266067T_0 + 1.130318T_1 + 0.2714953T_2 + 0.04433685T_3$

10.

$$r_T(x) = \frac{0.7584655T_0(x) - 0.09164152T_1(x) - 0.08639383T_2(x)}{T_0(x) - 0.9342827T_1(x) + 0.1504103T_2(x)}$$

Exercise Set 8.5 (PAGE 470)

2. a) 0.0   b) 5.143552   c) 4.0   d) 7.786191×10^{-3}

4. $a_0 = 0.1240293$
   $a_1 = -0.8600803$
   $a_2 = 2.549330$
   $a_3 = -0.6409933$
   $b_1 = -0.8321197$
   $b_2 = -0.6695062$
   The error is 107.913.
   The approximation in Exercise 4 is better because, in this case,

   $$\sum_{j=0}^{n} \left(f(\xi_j) - S_3(\xi_j) \right)^2 = 397.3678,$$

   whereas the approximation in Exercise 3 has

   $$\sum_{j=0}^{n} (f(\xi_j) - S_3(\xi_j))^2 = 569.3589.$$

6. a) For S_2:
   $a_0 = 2.739196$
   $a_1 = 0.09962390$
   $a_2 = -0.1771876$
   $b_1 = 1.053530$
   The error is 5.064436.
   For S_3:
   $a_0 = 2.739196$
   $a_1 = 0.09962390$
   $a_2 = -0.1771876$
   $a_3 = 0.1921922$
   $b_1 = 1.053530$
   The error is 2.70301.
   b) $S_2 = -0.8756590 + 0.5922466x + 0.2686725x^2$ has error 8.3747×10^{-4}.
   $S_3 = -0.5772120 + 0.07216164x + 0.5520643x^2 - 0.04875558x^3$ has error 1.502×10^{-5}.

91

8. a) $a_0 = 0.4205545$

$a_1 = -0.09802618$

$a_2 = 0.002905236$

$a_3 = 0.01328013$

$a_4 = -0.01884123$

$b_1 = 0.2398368$

$b_2 = -0.1290849$

$b_3 = 0.08577849$

b) 0.2102773   c) 0.2232443

Exercise Set 8.6 (PAGE 480)

2. Parts a) and b) give the same answer:

$a_0 = -9.252754$

$a_1 = 6.679518$

$a_2 = -3.701102$

$a_3 = 3.190086$

$a_4 = -3.084251$

$b_1 = 5.956833$

$b_2 = -2.467401$

$b_3 = 1.022031$

4. a) $a_0 = 0.3471000$

$a_1 = -0.02475498$

$a_2 = -0.0697570$

$a_3 = 0.08468317$

$a_4 = -0.08772957$

$b_1 = 0.2268260$

$b_2 = -0.1021640$

$b_3 = 0.04284648$

b) 0.1735500   c) 0.2232443

6. The b_k terms are all zero. The a_k are given in the following table.

k	a_k
0	−4.01287586
1	3.80276903
2	−2.23519870
3	0.63810403
4	−0.31550821
5	0.19408145
6	−0.13464491
7	0.10100593
8	−0.08015708
9	0.06643598
10	−0.05704353
11	0.05046675
12	−0.04583431
13	0.04262318
14	−0.04051395
15	0.03931584
16	−0.03892713

8. From Eq. (8.50)

$$c_k = \sum_{j=0}^{2m-1} y_j e^{\frac{\pi i j k}{m}} = \sum_{j=0}^{2m-1} y_j (\zeta)^{jk} = \sum_{j=0}^{2m-1} y_j (\zeta^k)^j.$$

Thus,

$$c_k = (1, \zeta^k, \zeta^{2k}, \ldots, \zeta^{(2m-1)k})^t \begin{bmatrix} y_0 \\ y_1 \\ \vdots \\ y_{2m-1} \end{bmatrix},$$

and the result follows.

9.

$$\begin{bmatrix} c_0 \\ c_1 \\ c_2 \\ c_3 \end{bmatrix} = \begin{bmatrix} 1 & 1 & 1 & 1 \\ 1 & \zeta & \zeta^2 & \zeta^3 \\ 1 & \zeta^2 & \zeta^4 & \zeta^6 \\ 1 & \zeta^3 & \zeta^6 & \zeta^9 \end{bmatrix} \begin{bmatrix} y_0 \\ y_1 \\ y_2 \\ y_3 \end{bmatrix}$$

$$
\begin{bmatrix} c_0 \\ c_1 \\ c_2 \\ c_3 \end{bmatrix} = \begin{bmatrix} 1 & 1 & 1 & 1 \\ 1 & \zeta^2 & \zeta^4 & \zeta^6 \\ 1 & \zeta & \zeta^2 & \zeta^3 \\ 1 & \zeta^3 & \zeta^6 & \zeta^9 \end{bmatrix} \begin{bmatrix} y_0 \\ y_1 \\ y_2 \\ y_3 \end{bmatrix} = \begin{bmatrix} 1 & 1 & 1 & 1 \\ 1 & \zeta^2 & 1 & \zeta^2 \\ 1 & \zeta & \zeta^2 & \zeta^3 \\ 1 & \zeta^3 & \zeta^2 & \zeta \end{bmatrix} \begin{bmatrix} y_0 \\ y_1 \\ y_2 \\ y_3 \end{bmatrix},
$$

since $\zeta^4 = 1, \zeta^5 = \zeta, \zeta^6 = \zeta^2, \zeta^7 = \zeta^3, \zeta^8 = 1, \zeta^9 = \zeta$. Hence

$$
\begin{bmatrix} c_0 \\ c_2 \\ c_1 \\ c_3 \end{bmatrix} = \begin{bmatrix} 1 & 1 & 0 & 0 \\ 1 & \zeta^2 & 0 & 0 \\ 0 & 0 & 1 & \zeta \\ 0 & 0 & 1 & \zeta^3 \end{bmatrix} \begin{bmatrix} 1 & 0 & 1 & 0 \\ 0 & 1 & 0 & 1 \\ 1 & 0 & \zeta^2 & 0 \\ 0 & 1 & 0 & \zeta^2 \end{bmatrix} \begin{bmatrix} y_0 \\ y_1 \\ y_2 \\ y_3 \end{bmatrix}.
$$

This shows the subdivision of the sums, since

$$
\begin{bmatrix} c_0 \\ c_2 \\ c_1 \\ c_3 \end{bmatrix} = \begin{bmatrix} 1 & 1 & 0 & 0 \\ 1 & \zeta^2 & 0 & 0 \\ 0 & 0 & 1 & \zeta \\ 0 & 0 & 1 & \zeta^3 \end{bmatrix} \begin{bmatrix} y_0 + y_2 \\ y_1 + y_3 \\ y_0 + \zeta^2 y_2 \\ y_1 + \zeta^2 y_3 \end{bmatrix}.
$$

10. $\mathbf{c} = A\mathbf{d}$, $\mathbf{d} = B\mathbf{e}$, $\mathbf{e} = C\mathbf{f}$, and $\mathbf{f} = D\mathbf{y}$, where

$$
A = \begin{bmatrix}
1 & 1 & 0 & 0 & 0 & 0 & 0 & 0 \\
0 & 0 & 1 & 1 & 0 & 0 & 0 & 0 \\
0 & 0 & 0 & 0 & 1 & 1 & 0 & 0 \\
0 & 0 & 0 & 0 & 0 & 0 & 1 & 1 \\
1 & -1 & 0 & 0 & 0 & 0 & 0 & 0 \\
0 & 0 & 1 & -1 & 0 & 0 & 0 & 0 \\
0 & 0 & 0 & 0 & 1 & -1 & 0 & 0 \\
0 & 0 & 0 & 0 & 0 & 0 & 1 & -1
\end{bmatrix},
$$

$$
B = \begin{bmatrix}
1 & 1 & 0 & 0 & 0 & 0 & 0 & 0 \\
0 & 0 & -i & -i & 0 & 0 & 0 & 0 \\
0 & 0 & 0 & 0 & 1 & 1 & 0 & 0 \\
0 & 0 & 0 & 0 & 0 & 0 & -i & -i \\
1 & -1 & 0 & 0 & 0 & 0 & 0 & 0 \\
0 & 0 & 1 & -1 & 0 & 0 & 0 & 0 \\
0 & 0 & 0 & 0 & 1 & -1 & 0 & 0 \\
0 & 0 & 0 & 0 & 0 & 0 & 1 & -1
\end{bmatrix},
$$

$$
C = \begin{bmatrix}
1 & 1 & 0 & 0 & 0 & 0 & 0 & 0 \\
0 & 0 & -i & -i & 0 & 0 & 0 & 0 \\
0 & 0 & 0 & 0 & \frac{-i+1}{\sqrt{2}} & \frac{-i+1}{\sqrt{2}} & 0 & 0 \\
0 & 0 & 0 & 0 & 0 & 0 & \frac{-i+1}{\sqrt{2}} & \frac{-i+1}{\sqrt{2}} \\
1 & -1 & 0 & 0 & 0 & 0 & 0 & 0 \\
0 & 0 & 1 & -1 & 0 & 0 & 0 & 0 \\
0 & 0 & 0 & 0 & 1 & -1 & 0 & 0 \\
0 & 0 & 0 & 0 & 0 & 0 & 1 & -1
\end{bmatrix},
$$

and

$$D = \begin{bmatrix} 1 & 0 & 0 & 0 & 0 & 0 & 0 & 0 \\ 0 & 0 & 0 & 0 & 1 & 0 & 0 & 0 \\ 0 & 0 & i & 0 & 0 & 0 & 0 & 0 \\ 0 & 0 & 0 & 0 & 0 & 0 & i & 0 \\ 0 & \frac{i-1}{\sqrt{2}} & 0 & 0 & 0 & 0 & 0 & 0 \\ 0 & 0 & 0 & 0 & 0 & \frac{i-1}{\sqrt{2}} & 0 & 0 \\ 0 & 0 & 0 & \frac{-(i+1)}{\sqrt{2}} & 0 & 0 & 0 & 0 \\ 0 & 0 & 0 & 0 & 0 & 0 & 0 & -\frac{(i+1)}{\sqrt{2}} \end{bmatrix}.$$

Note that $\mathbf{c} = ABCD\mathbf{y}$, which would give Eq. (8.50) if expanded.

Chapter 9
Approximating Eigenvalues

Exercise Set 9.1 (PAGE 490)

2. a) Only 1(d).

b)

$$P = \begin{bmatrix} 0 & \frac{\sqrt{2}}{2} & \frac{\sqrt{2}}{2} \\ 1 & 0 & 0 \\ 0 & \frac{\sqrt{2}}{2} & -\frac{\sqrt{2}}{2} \end{bmatrix} \quad \text{and} \quad D = \begin{bmatrix} 2 & 0 & 0 \\ 0 & 3 & 0 \\ 0 & 0 & 1 \end{bmatrix}.$$

4. Let $\mathbf{w} = (w_1, w_2, w_3)^t, \mathbf{x} = (x_1, x_2, x_3)^t, \mathbf{y} = (y_1, y_2, y_3)^t$ and $\mathbf{z} = (z_1, z_2, z_3)^t$ be in $\mathbf{R}^3$. Suppose $\mathbf{w}, \mathbf{x},$ and $\mathbf{y}$ are linearly independent. Consider the linear system

$$\begin{bmatrix} w_1 & x_1 & y_1 \\ w_2 & x_2 & y_2 \\ w_3 & x_3 & y_3 \end{bmatrix} \begin{bmatrix} a \\ b \\ c \end{bmatrix} = \begin{bmatrix} 0 \\ 0 \\ 0 \end{bmatrix}.$$

Since $\mathbf{w}, \mathbf{x},$ and $\mathbf{y}$ are linearly independent, the only solution is $a = b = c = 0$. Thus, the determinant of the matrix is nonzero. Hence, the linear system

$$\begin{bmatrix} w_1 & x_1 & y_1 \\ w_2 & x_2 & y_2 \\ w_3 & x_3 & y_3 \end{bmatrix} \begin{bmatrix} a \\ b \\ c \end{bmatrix} = \begin{bmatrix} z_1 \\ z_2 \\ z_3 \end{bmatrix}$$

has a unique solution. Thus, $a\mathbf{w} + b\mathbf{x} + c\mathbf{y} - \mathbf{z} = 0$, so $\{\mathbf{w}, \mathbf{x}, \mathbf{y}, \mathbf{z}\}$ is linearly dependent. If $\{\mathbf{w}, \mathbf{x}, \mathbf{y}\}$ is linearly dependent, then $\{\mathbf{w}, \mathbf{x}, \mathbf{y}, \mathbf{z}\}$ also is linearly dependent.

6. Let $\mathbf{P}_1, \ldots, \mathbf{P}_n$ denote the columns of P. Since $P^t P = I$,

$$\begin{bmatrix} \mathbf{P}_1^t \\ \mathbf{P}_2^t \\ \mathbf{P}_3^t \\ \vdots \\ \mathbf{P}_n^t \end{bmatrix} [\mathbf{P}_1, \mathbf{P}_2, \ldots, \mathbf{P}_n] = I.$$

Thus, $(P^t P)_{i,j} = \mathbf{P}_i^t \mathbf{P}_j = (I)_{i,j}$ and $\{\mathbf{P}_i\}$ is an orthonormal set. Hence

$$\|P\|_2 = (\rho(P^t P))^{\frac{1}{2}} = (\rho(I))^{\frac{1}{2}} = 1$$

and similarly $\|P^t\|_2 = 1$.

8. Let $A\mathbf{x}^{(i)} = \lambda_i \mathbf{x}^{(i)}$ for $i = 1, 2, \ldots, n$ where the λ_i are distinct. Suppose $\{\mathbf{x}^{(i)}\}_{i=1}^k$ is the largest linearly independent set of eigenvectors of A where $1 \leq k < n$. (Note that a re-indexing may be necessary for the preceding statement to hold.)

Since $\{\mathbf{x}^{(i)}\}_{i=1}^{k+1}$ is linearly dependent, there exist numbers $c_1, \ldots, c_{k+1}$, not all zero, with

$$c_1 \mathbf{x}^{(1)} + \ldots + c_k \mathbf{x}^{(k)} + c_{k+1} \mathbf{x}^{(k+1)} = 0.$$

Since $\{\mathbf{x}^{(i)}\}_{i=1}^{k}$ is linearly independent, $c_{k+1} \neq 0$. Multiplying by A gives

$$c_1 \lambda_1 \mathbf{x}^{(1)} + \ldots + c_k \lambda_k \mathbf{x}^{(k)} + c_{k+1} \lambda_{k+1} \mathbf{x}^{(k+1)} = 0.$$

Thus,

$$\frac{c_1(\lambda_{k+1} - \lambda_1)}{c_{k+1}} \mathbf{x}^{(1)} + \ldots + \frac{c_k(\lambda_{k+1} - \lambda_k)}{c_{k+1}} \mathbf{x}^{(k)} = 0.$$

But $\{\mathbf{x}^{(i)}\}_{i=1}^{k}$ is linearly independent and $\mathbf{x}^{(k+1)} \neq 0$, so $\lambda_{k+1} = \lambda_i$ for some $1 \leq i \leq k$.

10. a) Let μ be an eigenvalue of A. Since A is symmetric, μ is real and Theorem 9.14 gives $0 \leq \mu \leq 4$. The eigenvalues of $A - 4I$ are of the form $\mu - 4$. Thus,

$$\rho(A - 4I) = \max|\mu - 4| = \max(4 - \mu) = 4 - \min\mu = 4 - \lambda = |\lambda - 4|.$$

b) The eigenvalues of $A - 4I$ are $-3.618034, -2.618034, -1.381966, -0.381966$ so $\rho(A - 4I) = 3.618034$ and $\lambda = 0.381966$. An eigenvector is $(0.618034, 1, 1, 0.618034)^t$.

c) As in part a), $0 \leq \mu \leq 6$ so $|\lambda - 6| = \rho(B - 6I)$.

d) The eigenvalues of $B - 6I$ are $-5.2360673, -4, -2, -0.76393202$. $\rho(B - 6I) = 5.2360673$ and $\lambda = 0.7639327$. An eigenvector is $(0.61803395, 1, 1, 0.6180395)^t$.

Exercise Set 9.2 (PAGE 505)

2. a) With $\mathbf{x}^{(0)} = (1, 0, 0)^t$, $q = 1$ and $A - qI$ is singular. So $\lambda = 1$ is an eigenvalue.

b) $\mathbf{x}^{(3)} = (0.2777778, 0.7777778, 1)^t$, $\mu^{(3)} = -0.3571429$, and $\lambda \simeq -1.8$

c) $\mathbf{x}^{(3)} = (0.7024682, 1, -0.3164256, 0.5615173)^t$, $\mu^{(3)} = 0.6568335$, and $\lambda \simeq 5.772456$

d) $\mathbf{x}^{(1)} = (1, 0.9999997, 1, 0.9999997)^t$, $\mu^{(1)} = -5.592403 \times 10^6$, and $\lambda \simeq 3.999999$ with convergence ocurring.

4. a) Change the Inverse Power Method Algorithm 9.3 as follows:

Replace STEPS 3 and 4 with

set

$$\mathbf{x} = \frac{\mathbf{x}}{||\mathbf{x}||_2}$$

Replace STEP 7 with

$$\text{set } \mu = \mathbf{x}^t \mathbf{y}.$$

Replace STEPS 8 and 9 with

set

$$\text{ERR} = \left\| \mathbf{x} - \frac{\mathbf{y}}{\|\mathbf{y}\|_2} \right\|_2$$

$$\mathbf{x} = \frac{\mathbf{y}}{\|\mathbf{y}\|_2}$$

b) (3a) $q = 2$ is an eigenvalue so $A - 2I$ is singular.

(3b) $\lambda = 4.965259; \mathbf{x} = (-0.4717782, 0.05369533, 0.8800808)^t$

(3c) $\lambda = 2.467290; \mathbf{x} = (0.6270597, -0.6967330, 0.2090199, -0.2786932)^t$

(3d) $q = 4$ is an eigenvalue so $A - 4I$ is singular.

6. (1a) $\mu^{(3)} = 5, \mathbf{x}^{(5)} = (-0.26, 1, -0.24)^t$
   (1b) $\mu^{(7)} = 5.124886, \mathbf{x}^{(7)} = (-0.2424496, 1, -0.3199686)^t$
   (1c) $\mu^{(13)} = 5.236072, \mathbf{x}^{(13)} = (1, 0.6167300, 0.1189088, 0.4994857)^t$
   (1d) $\mu^{(2)} = 4, \mathbf{x} = (1, 1, 1, 1)^t$

8. a) With $\mathbf{x}^{(0)} = (1, 0, 0)^t, q = 2$ which is an eigenvalue. So $A - 2I$ is singular.
   With $\mathbf{x}^{(0)} = (1, 1, 1)^t, \mathbf{x}^{(4)} = (0.7071070, 1, 0.7071070)^t, \lambda = 0.5857861.$
   b) With $\mathbf{x}^{(0)} = (0, 1, 0)^t, q = 4.75, \mathbf{x}^{(6)} = (-0.5499681, 0.05937225, 1)^t, \lambda = 4.961707.$
   c) With $\mathbf{x}^{(0)} = (0, 1, 0, 0)^t, q = 3, \mathbf{x}^{(15)} = (-0.9449723, 1, -0.4308172, 0.5690715)^t,$
   $\lambda = 2.485845.$
   d) With $\mathbf{x}^{(0)} = (1, 0, 0, 0)^t, q = 4$ is an eigenvalue. So $A - 4I$ is singular.

Using the results of Exercise 7 for input:
   (3a) $\lambda_2 = 1.999914, \mathbf{x}^{(2)} = (1.414265, -0.00003575585, -1.414265)^t$
   (3b) $\lambda_2 = 4.961943, \mathbf{x}^{(2)} = (1.275705, -0.1375766, -2.319548)^t$
   (3c) $\lambda_2 = 4.427366, \mathbf{x}^{(2)} = (-2.483645, -1.460986, -0.3976560, -1.858642)^t$
   (3d) $\lambda_2 = 3.620882, \mathbf{x}^{(2)} = (0.7236983, -1.171112, 1.170154, -0.2759610)^t$

10. a) Let c be given with the same sign as λ_1, and with $|c| > |\lambda_1|$. The eigenvalues of $cI - A$
   are $c - \lambda_i$.
   Suppose $\lambda_1 \geq \ldots \geq \lambda_{n-1} > \lambda_n$ with $|\lambda_1| > |\lambda_n|$. Then $c - \lambda_1 \leq \ldots \leq c - \lambda_{n-1} < c - \lambda_n$.
   Since c and λ_1 have the same sign

$$|c - \lambda_1| = |c| - |\lambda_1| < |c| - |\lambda_n| \leq |c - \lambda_n|$$

and $c - \lambda_n$ is the dominant eigenvalue of $cI - A$.

If $\lambda_1 \leq \ldots \leq \lambda_{n-1} < \lambda_n$ and $|\lambda_1| > |\lambda_n|$, then

$$c - \lambda_1 \geq \ldots \geq c - \lambda_{n-1} > c - \lambda_n.$$

But again,

$$|c - \lambda_1| = |c| - |\lambda_1| < |c| - |\lambda_n| \leq |c - \lambda_n|$$

and $c - \lambda_n$ is the dominant eigenvalue of $cI - A$.

b) (1a) With $c = 6, \lambda_3 = 0$ and $\mathbf{x}^{(3)} = (1, 1, 1)^t$.

(1b) With $c = 6, \lambda_3 = 0.2384204$ and $\mathbf{x}^{(3)} = (1, 0.7615742, 0.4323470)^t$.

(1c) With $c = 6, \lambda_4 = 0.7639437$ and $\mathbf{x}^{(4)} = (-0.4721550, 0.7639028, 1, -0.2359938)^t$.

(1d) With $c = 10, \lambda_4 = 2.0000359$ and $\mathbf{x}^{(6)} = (1, 0.9999999, -0.9999731, -0.9999732)^t$.

11. c) (1a) Using $\mathbf{x}^{(0)} = (-4, -2, 0)^t$ gives $\lambda_2 = 1$ and $\mathbf{x}^{(2)} = (1, 0, -1)^t$.
   (1b) Using $\mathbf{x}^{(0)} = (-4.12488543, -2, 0)$ gives $\lambda_2 = 1.636682$ and
   $\mathbf{x}^{(2)} = (-0.5706754, 0.3633349, 1)^t$.
   (1c) The method fails to obtain λ_2.
   (1d) Using $\mathbf{x}^{(0)} = (-1.23606796, 1, -1, 0)^t$ gives $\lambda_2 = 5.9997153$
   and $\mathbf{x}^{(2)} = (1, 0.9995254, 1, 0.9995254)^t$.

   d) The proof is an extension of parts a) and b).

12. a) Let $\{\mathbf{v}^{(1)}, \ldots, \mathbf{v}^{(n)}\}$ be an orthogonal set of eigenvectors of A. Then

$$B\mathbf{v}^{(1)} = A\mathbf{v}^{(1)} - \frac{\lambda_1}{(\mathbf{v}^{(1)})^t(\mathbf{v}^{(1)})}\mathbf{v}^{(1)}(\mathbf{v}^{(1)})^t\mathbf{v}^{(1)} = A\mathbf{v}^{(1)} - \lambda_1\mathbf{v}^{(1)} = 0 \cdot \mathbf{v}^{(1)} = 0$$

and, for $i \neq 1$,

$$B\mathbf{v}^{(i)} = A\mathbf{v}^{(i)} - \frac{\lambda_1}{(\mathbf{v}^{(1)})^t(\mathbf{v}^{(1)})}\mathbf{v}^{(1)}(\mathbf{v}^{(1)})^t\mathbf{v}^{(i)} = A\mathbf{v}^{(i)} = \lambda_i\mathbf{v}^{(i)}.$$

b) (3a) $\lambda_2 = 2, \mathbf{x}^{(2)} = (1, 0, -1)^t$
   (3b) $\lambda_2 = 4.96169914, \mathbf{x}^{(2)} = (-0.54996974, 0.05936537, 1)^t$
   (3c) $\lambda_2 = 4.42800686, \mathbf{x}^{(2)} = (1, 0.58813965, 0.16013279, 0.79827238)^t$.
   (3d) $\lambda_2 = 3.618034, \mathbf{x}^{(2)} = (0.61803403, -0.99999998, 1, -0.23606796)^t$

14. a) $|\lambda| \leq 6$ for all eigenvalues λ.
   b) $\lambda_1 = 0.6982681, \mathbf{x} = (1, 0.71606, 0.25638, 0.04602)^t$
   d) $P(\lambda) = \lambda^4 - \frac{1}{4}\lambda - \frac{1}{16}$;
   $\lambda_1 = 0.6976684972,$
   $\lambda_2 = -0.237313308,$

$$\lambda_3 = -0.2301775942 + 0.56965884i,$$
$$\lambda_4 = -0.2301775942 - 0.56965884i.$$

e) The beetle population should approach zero since A is convergent.

Exercise Set 9.3 (PAGE 514)

2. a)

$$\begin{bmatrix} 4.0000000 & 1.4142136 & 0 & 0 \\ 1.4142136 & 4.0000000 & 1.4142136 & 0 \\ 0 & 1.4142136 & 4.0000000 & 0 \\ 0 & 0 & 0 & 4.0000000 \end{bmatrix}$$

b)

$$\begin{bmatrix} 5.0000000 & 2.5495098 & 0 & 0 \\ 2.5495098 & 6.38461538 & 2.1407569 & 0 \\ 0 & 2.1407569 & 4.2700005 & 0.6912809 \\ 0 & 0 & 0.6912809 & 4.345384 \end{bmatrix}$$

c)

$$\begin{bmatrix} 8.0000000 & -2.3048861 & 0 & 0 & 0 \\ -2.3048861 & 5.9294118 & 1.5022590 & 0 & 0 \\ 0 & 1.5022590 & 1.7714975 & -4.8901511 & 0 \\ 0 & 0 & -4.8901511 & -0.4361218 & -1.0898884 \\ 0 & 0 & 0 & -1.0898884 & 4.7352125 \end{bmatrix}$$

d)

$$\begin{bmatrix} 2.0000000 & 1.4142136 & 0 & 0 & 0 \\ 1.4142136 & 3.5000000 & 0.8660254 & 0 & 0 \\ 0 & 0.8660254 & 7.8333333 & 4.7140452 & 0 \\ 0 & 0 & 4.7140452 & 6.6666667 & 1.7320508 \\ 0 & 0 & 0 & 1.7320508 & 6.0000000 \end{bmatrix}$$

4. (1a)

$$\begin{bmatrix} 12.0 & -10.7 & 0.0 \\ -10.7 & 3.92 & 5.31 \\ 0.0 & 5.31 & 7.12 \end{bmatrix} ; \quad 0.1631$$

(1b)

$$\begin{bmatrix} 2.00 & 1.41 & 0.0 \\ 1.41 & 1.01 & 0.00500 \\ 0.0 & 0.00500 & 3.00 \end{bmatrix} ; \quad 0.01921$$

(1c)

$$\begin{bmatrix} 2.00 & 1.00 & 0.0 \\ 1.00 & -1.00 & -2.00 \\ 0.0 & -2.00 & 3.00 \end{bmatrix} ; \quad 0.0$$

(1d)

$$\begin{bmatrix} 4.75 & -2.26 & 0.0 \\ -2.26 & 4.48 & -1.22 \\ 0.0 & -1.22 & 5.03 \end{bmatrix} ; \quad 0.008724$$

Exercise Set 9.4 (PAGE 524)

2. a) $3.9115033, 2.1294613, -2.0409646$

   b) $1.2087122, 5.7912878, 3.0000000$

   c) $6.0000000, 2.0000000, 4.0000000, 7.4641016, 0.5358984$

   d) $4.0274350, 2.0707128, 3.7275564, 5.7839956, 0.8903002$

4. a) $(-0.7071067, 1, -0.7071067)^t, (1, 0, -1)^t, (0.7071068, 1, 0.7071068)^t$

   b) $(0.1741299, -0.5343539, 1)^t, (0.4261735, 1, 0.4601443)^t,$
      $(1, -0.2777544, -0.3225491)^t$

   c) $(-0.1520150, -0.3008950, -0.05155956, 1)^t, (0.3627966, 1, 0.7459807, 0.3945081)^t,$
      $(1, 0.09528962, -0.6907921, 0.1450703)^t,$ and $(0.8029403, -0.9884448, 1, -0.1237995)^t$

   d) $(-0.2172064, -0.1253620, 0.3108802, 1)^t, (1, 0.9718785, -0.4885653, 0.4909285)^t,$
      $(1, -0.9482070, 0.1976567, 0.03688974)^t,$ and $(0.1978041, 0.4086314, 1, -0.2166890)^t$

6. Let

$$P = \begin{bmatrix} 1 & & & & & & \\ & \ddots & & & & & \\ & & \cos\theta & \cdots & -\sin\theta & & \\ & & \vdots & \ddots & \vdots & & \\ & & \sin\theta & \cdots & \cos\theta & & \\ & & & & & \ddots & \\ & & & & & & 1 \end{bmatrix} \begin{matrix} \\ \\ \text{row } j \\ \\ \text{row } i \\ \\ \end{matrix}$$

$$\text{column } j \quad \text{column } i$$

be a rotation matrix. For any $n \times n$ matrix A,

$$(AP)_{p,q} = \sum_{k=1}^{n} a_{pk} P_{kq}.$$

If $q \neq i, j$, then $P_{kq} = 0$ unless $k = q$. Thus, $(AP)_{pq} = a_{pq}$.

If $q = j$, then

$$(AP)_{p,j} = a_{pj} P_{jj} + a_{pi} P_{ij} = a_{pj} \cos\theta + a_{pi} \sin\theta.$$

101

If $q = i$, then

$$(AP)_{p,i} = a_{pj}P_{ji} + a_{pi}P_{ii} = -a_{pj}\sin\theta + a_{pi}\cos\theta.$$

Similiarly, $(PA)_{p,q} = \sum_{k=1}^{n} P_{p,k}a_{k,q}$. If $p \neq i,j$, then $P_{p,k} = 0$ unless $p = k$.
Thus, $(PA)_{p,q} = P_{pp}a_{pq} = a_{pq}$.
If $p = i$, then

$$(PA)_{i,q} = P_{i,j}a_{jq} + P_{ii}a_{iq} = a_{jq}\sin\theta + a_{iq}\cos\theta.$$

If $p = j$, then

$$(PA)_{j,q} = P_{jj}a_{jq} + P_{ji}a_{iq} = a_{jq}\cos\theta - a_{iq}\sin\theta.$$

7. INPUT: dimension n, matrix $A = (a_{ij})$, tolerance TOL,
    maximum number of iterations N.

OUTPUT: eigenvalues $\lambda_1, \ldots, \lambda_n$ of A or a message that the number of iterations was exceeded.

STEP 1 Set $FLAG = 1; k1 = 1$.

STEP 2 While $(FLAG = 1)$ do STEPS 3 – 10

    STEP 3 For $i = 2, \ldots, n$ do STEPS 4 – 8.

        STEP 4 For $j = 1, \ldots, i-1$ do STEPS 5 – 8.

            STEP 5 If $a_{ii} = a_{jj}$ then set
$$CO = 0.5\sqrt{2};$$
$$SI = CO$$
else set
$$b = |a_{ii} - a_{jj}|;$$
$$c = 2a_{ij}\,\text{sign}(a_{ii} - a_{jj});$$
$$CO = (0.5(1 + b/(c^2 + b^2)^{\frac{1}{2}})^{\frac{1}{2}});$$
$$SI = 0.5c/(CO(c^2 + b^2)^{\frac{1}{2}}).$$

            STEP 6 For $k = 1, \ldots, n$ set
$$x = a_{k,j};$$
$$y = a_{k,i};$$
$$a_{k,j} = CO \cdot x + SI \cdot y;$$
$$a_{k,i} = CO \cdot y + SI \cdot x.$$

            STEP 7 For $k = 1, \ldots, n$ set
$$x = a_{j,k};$$
$$y = a_{i,k};$$
$$a_{j,k} = CO \cdot x - SI \cdot y;$$
$$a_{i,k} = SI \cdot x + CO \cdot y.$$

            STEP 8 Set $a_{i,j} = 0;\quad a_{j,i} = 0$.

102

STEP 9 Set

$$s = \sum_{i=1}^{n} \sum_{\substack{j=1 \\ j \neq i}}^{n} |a_{ij}|.$$

STEP 10 If $s < TOL$ then

    for $i = 1, ..., n$

     set $\lambda_i = a_{ii}$;

    OUTPUT $(\lambda_1, ..., \lambda_n)$;

    set FLAG = 0.

   else set $k1 = k1 + 1$;

    if $k1 > N$ then set FLAG = 0.

STEP 11 If $k1 > N$ then

    OUTPUT ('Maximum number of iterations exceeded');

    STOP.

8. Using tolerance 10^{-5} in all cases:

 (1a) $3.414214, 0.5857864, 2.0000000$; 3 iterations

 (1b) $2.722246, 5.346462, -0.06870782$; 3 iterations

 (1c) $0.1922421, 1.189091, 0.5238224, 0.9948440$; 3 iterations

 (1d) $6.844621, 1.084364, -2.197517, 2.268531$; 3 iterations

Chapter 10
Numerical Solutions of Nonlinear Systems of Equations

Exercise Set 10.1 (PAGE 534)

2. $F(x_1, x_2) = \left(1, \frac{1}{|x_1 - 1| + |x_2|}\right)^t$

4. a) The solutions are near $(-1.5, -10.5)$ and $(2, 11)$.

   b) Use $G_1 = (-0.5 + \sqrt{2x_2 - 17.75}, 6 + \sqrt{25 - (x_1 - 1)^2})^t$
   and $G_2 = (-0.5 - \sqrt{2x_2 - 17.75}, 6 + \sqrt{25 - (x_1 - 1)^2})^t$.

   c) For G_1 with $x^{(0)} = (2, 11)^t$ we have $x^{(9)} = (1.603990, 10.96338)^t$ and for G_2 with $x^{(0)} = (-1.5, -10.5)$, we have $x^{(34)} = (-2.000003, 9.999996)^t$.

6. a) $G = (x_2/\sqrt{5}, 0.25(\sin x_1 + \cos x_2))^t$, and $D = \{(x_1, x_2) | 0 \le x_1, x_1 \le 1\}$.

   b) With $x^{(0)} = (\frac{1}{2}, \frac{1}{2})^t, x^{(10)} = (0.1212440, 0.2711065)^t$.

   c) With $x^{(0)} = (\frac{1}{2}, \frac{1}{2})^t, x^{(5)} = (0.1212421, 0.2711052)^t$.

8. a) With $x^{(0)} = (1, 1, 1)^t, x^{(3)} = (0.5000000, 0, -0.5235988)^t$.

   b) With $x^{(0)} = (1, 1, 1)^t, x^{(4)} = (1.036400, 1.085707, 0.9311914)^t$.

   c) With $x^{(0)} = (0.05, 0.2, 0.8)^t, x^{(3)} = (0, 0.1000000, 1.0000000)^t$.

   d) With $x^{(0)} = (0, 0, 0)^t, x^{(4)} = (0.4981447, -0.1996059, -0.5288260)^t$.

10. Let $F(x) = (f_1(x), ..., f_n(x))^t$. Suppose F is continuous at x_0. By Definitions 10.3 and 10.4,

$$\lim_{x \to x_0} f_i(x) = f_i(x_0)$$

for each $i = 1, ..., n$.

Given $\epsilon > 0$ there exists $\delta_i > 0$ such that

$$|f_i(x) - f_i(x_0)| < \epsilon$$

whenever $0 < \|x - x_0\| < \delta_i$ and $x \in D$.
Let $\delta = \min_{1 \le i \le n} \delta_i$. If $0 < \|x - x_0\| < \delta$, then $|f_i(x) - f_i(x_0)| < \epsilon$ for each $i = 1, ..., n$ whenever $x \in D$.
Then

$$\|F(x) - F(x_0)\| < \epsilon$$

whenever $\|x - x_0\| < \delta$ and $x \in D$.

By the equivalence of vector norms, the result holds for all vector norms by suitably adjusting δ.

For the converse, let $\epsilon > 0$ be given. Then there is a $\delta > 0$ such that

$$\|\mathbf{F}(\mathbf{x}) - \mathbf{F}(\mathbf{x}_0)\| < \epsilon$$

whenever $\mathbf{x} \in D$ and $\|\mathbf{x} - \mathbf{x}_0\| < \delta$. By the equivalence of vector norms, $\delta' > 0$ can be found with

$$\|f_i(\mathbf{x}) - f_i(\mathbf{x}_0)\| < \epsilon$$

whenever $\mathbf{x} \in D$ and $\|\mathbf{x} - \mathbf{x}_0\| < \delta'$.

Thus, $\lim_{\mathbf{x} \to \mathbf{x}_0} f_i(\mathbf{x}) = f_i(\mathbf{x}_0)$ for $i = 1, ..., n$. Since $\mathbf{F}(\mathbf{x}_0)$ is defined, $\mathbf{F}$ is continuous at $\mathbf{x}_0$ by Definition 10.4.

Exercise Set 10.2 (PAGE 541)

2. a) With $\mathbf{x}^{(0)} = (1, 1, 1)^t, \mathbf{x}^{(8)} = (6, 1, -4)^t$.

   b) With $\mathbf{x}^{(0)} = (0.05, 0.2, 0.8)^t, \mathbf{x}^{(3)} = (0, 0.1, 1)^t$.

   c) With $\mathbf{x}^{(0)} = (1, 1, 1)^t, \mathbf{x}^{(3)} = (1.0364005, 1.085707, 0.9311914)^t$.

   d) With $\mathbf{x}^{(0)} = (0, 0, 0)^t, \mathbf{x}^{(2)} = (1, 2, 0)^t$.

4. a) With $\mathbf{x}^{(0)} = (1, 1, 1)^t, \mathbf{x}^{(9)} = (0.4999818, 0.01999927, -0.5231013)^t$.

   b) With $\mathbf{x}^{(0)} = (2, 0, 0, 0)^t, \mathbf{x}^{(18)} = (1.000006, 1.000006, 0.9999984, 0.9999984)^t$.

5.

$$J(\mathbf{x}) = \begin{pmatrix} c_{11} & c_{12} & \cdots & c_{1n} \\ c_{21} & c_{22} & \cdots & c_{2n} \\ \vdots & \vdots & \ddots & \vdots \\ c_{n1} & c_{n2} & \cdots & c_{nn} \end{pmatrix}$$

6. With $\theta_i^{(0)} = 1$ for each $i = 1, 2, \ldots, 20$ the following results are obtained:

i	$\theta_i^{(5)}$		i	$\theta_i^{(5)}$
1	0.14062		11	0.48348
2	0.19954		12	·0.50697
3	0.24522		13	0.52980
4	0.28413		14	0.55205
5	0.31878		15	0.57382
6	0.35045		16	0.59516
7	0.37990		17	0.61615
8	0.40763		18	0.63683
9	0.43398		19	0.65726
10	0.45920		20	0.67746

8. With $\mathbf{x}^{(0)} = (0.75, 1.25)^t, \mathbf{x}^{(14)} = (0.7501135, 1.184862)^t$.

Thus, $a = 0.7501135, b = 1.184862$, and the error is 19.796.

9. a)

$$\frac{\partial E}{\partial a} = 2 \sum_{i=1}^{n} \left(w_i y_i - \frac{a}{(x_i - b)^c} \right) = 0,$$

$$\frac{\partial E}{\partial b} = 2 \sum_{i=1}^{n} \left(w_i y_i - \frac{a}{(x_i - b)^c} \right) \left(\frac{-ac}{(x_i - b)^{c+1}} \right) \left(\frac{-1}{(x_i - b)^c} \right) = 0,$$

and

$$\frac{\partial E}{\partial c} = 2 \sum_{i=1}^{n} \left(w_i y_i - \frac{a}{(x_i - b)^c} \right) \ln(x_i - b) \left(\frac{-a}{(x_i - b)^c} \right) = 0.$$

Solving for a in the first equation and substituting into the second and third equations gives the linear system.

b) With $\mathbf{x}^{(0)} = (26.8, 8.3)^t = (b_0, c_0)^t$, we have $\mathbf{x}^{(7)} = (26.77021, 8.451831)^t$.

Thus, $a = 2.217952 \times 10^6$, $b = 26.77021$, $c = 8.451831$, and

$$\sum_{i=1}^{n} \left(w_i y_i - \frac{a}{(x_i - b)^c} \right)^2 = 0.7821139.$$

Exercise Set 10.3 (PAGE 550)

2. a) With $\mathbf{x}^{(0)} = (0.1, -0.1, 0.1)^t$, $\mathbf{x}^{(4)} = (0.4981447, -0.1996059, -0.5288260)^t$.

   b) With $\mathbf{x}^{(0)} = (0.4, 0.3, 0.2)^t$, $\mathbf{x}^{(20)} = (0.5291505, 0.4000003, 0.1000002)^t$.

   c) With $\mathbf{x}^{(0)} = (-0.3, -0.3, 4)^t$, $\mathbf{x}^{(7)} = (-0.4342585, -0.4342585, 5.302776)^t$.

   d) With $\mathbf{x}^{(0)} = (0.8, 0.8, 0.8, 1.4)^t$, $\mathbf{x}^{(17)} = (0.8688769, 0.8688769, 0.8688769, 1.524492)^t$.

4. a) With $\mathbf{x}^{(0)} = (1, 1, 1)^t$, $\mathbf{x}^{(13)} = (0.4999818, 0.01999835, -0.5231013)^t$.

   b) The method will likely fail for any choice of $\mathbf{x}^{(0)}$.

6. With $\mathbf{x}^{(0)} = (0.75, 1.25)^t$, $\mathbf{x}^{(10)} = (0.7500917, 1.184893)^t$.

   Thus, $a = 0.7500917, b = 1.184893$, and the error is 19.796.

8.

$$\left[A^{-1} - \frac{A^{-1}\mathbf{x}\mathbf{y}^t A^{-1}}{1 + \mathbf{y}^t A^{-1}\mathbf{x}}\right](A + \mathbf{x}\mathbf{y}^t) = A^{-1}A - \frac{A^{-1}\mathbf{x}\mathbf{y}^t A^{-1}A}{1 + \mathbf{y}^t A^{-1}\mathbf{x}} + A^{-1}\mathbf{x}\mathbf{y}^t - \frac{A^{-1}\mathbf{x}\mathbf{y}^t A^{-1}\mathbf{x}\mathbf{y}^t}{1 + \mathbf{y}^t A^{-1}\mathbf{x}}$$

$$= I - \frac{A^{-1}\mathbf{x}\mathbf{y}^t}{1 + \mathbf{y}^t A^{-1}\mathbf{x}} + A^{-1}\mathbf{x}\mathbf{y}^t - \frac{A^{-1}\mathbf{x}\mathbf{y}^t A^{-1}\mathbf{x}\mathbf{y}^t}{1 + \mathbf{y}^t A^{-1}\mathbf{x}}$$

$$= I - \frac{A^{-1}\mathbf{x}\mathbf{y}^t - A^{-1}\mathbf{x}\mathbf{y}^t - \mathbf{y}^t A^{-1}\mathbf{x}A^{-1}\mathbf{x}\mathbf{y}^t + A^{-1}\mathbf{x}\mathbf{y}^t A^{-1}\mathbf{x}\mathbf{y}^t}{1 + \mathbf{y}^t A^{-1}\mathbf{x}}$$

$$= I + \frac{\mathbf{y}^t A^{-1}\mathbf{x}A^{-1}\mathbf{x}\mathbf{y}^t - \mathbf{y}^t A^{-1}\mathbf{x}(A^{-1}\mathbf{x}\mathbf{y}^t)}{1 + \mathbf{y}^t A^{-1}\mathbf{x}}$$

$$= I$$

Exercise Set 10.4 (PAGE 557)

2. a) $\mathbf{x}^{(2)} = (1.0000000, 0.9999999)^t$

   b) $\mathbf{x}^{(3)} = (1.546946, 10.96999)^t$

   c) $\mathbf{x}^{(2)} = (0.5000000, 0.8660254)^t$

   d) $\mathbf{x}^{(4)} = (-0.2605993, 0.6225309)^t$

4. a) $\mathbf{x}^{(5)} = (1, 1, 1)^t$

   b) $\mathbf{x}^{(2)} = (0.000000, 1.000000, 1)^t$

   c) $\mathbf{x}^{(4)} = (1.036400, 1.085707, 0.9311914)^t$

   d) $\mathbf{x}^{(3)} = (0.4999818, 0.01999927, -0.5231013)^t$

6. a) We have

$$\alpha_1 = 0$$

$$g_1 = g(x_1, ..., x_n) = g(\mathbf{x}^{(0)}) = h(\alpha_1)$$

$$g_3 = g(\mathbf{x}^{(0)} - \alpha_3 \nabla g(\mathbf{x}^{(0)})) = h(\alpha_3)$$

$$g_2 = g(\mathbf{x}^{(0)} - \alpha_2 \nabla g(\mathbf{x}^{(0)})) = h(\alpha_2)$$

$$h_1 = \frac{(g_2 - g_1)}{(\alpha_2 - \alpha_1)} = g[\mathbf{x}^{(0)} - \alpha_1 \nabla g(\mathbf{x}^{(0)}), \mathbf{x}^{(0)} - \alpha_2 \nabla g(\mathbf{x}^{(0)})] = h[\alpha_1, \alpha_2]$$

$$h_2 = \frac{(g_3 - g_2)}{(\alpha_3 - \alpha_2)} = g[\mathbf{x}^{(0)} - \alpha_2 \nabla g(\mathbf{x}^{(0)}), \mathbf{x}^{(0)} - \alpha_3 \nabla g(\mathbf{x}^{(0)})] = h[\alpha_2, \alpha_3]$$

$$h_3 = \frac{(h_2 - h_1)}{(\alpha_3 - \alpha_1)}$$

$$= g[\mathbf{x}^{(0)} - \alpha_1 \nabla g(\mathbf{x}^{(0)}), \mathbf{x}^{(0)} - \alpha_2 \nabla g(\mathbf{x}^{(0)}), \mathbf{x}^{(0)} - \alpha_3 \nabla g(\mathbf{x}^{(0)})]$$

$$= h[\alpha_1, \alpha_2, \alpha_3].$$

The Newton divided-difference form of the second interpolating polynomial is

$$P(\alpha) = h[\alpha_1] + h[\alpha_1, \alpha_2](\alpha - \alpha_1) + h[\alpha_1, \alpha_2, \alpha_3](\alpha - \alpha_1)(\alpha - \alpha_2)$$

$$= g_1 + h_1(\alpha - \alpha_1) + h_3(\alpha - \alpha_1)(\alpha - \alpha_2)$$

$$= g_1 + h_1 \alpha + h_3 \alpha(\alpha - \alpha_2).$$

b) $P'(\alpha) = h_1 - \alpha_2 h_3 + 2h_3 \alpha$. So $P'(\alpha) = 0$ when $\alpha = 0.5(\alpha_2 - h_1/h_3)$.

Chapter 11
Boundary-Value Problems for Ordinary Differential Equations

Exercise Set 11.1 (Page 565)

2.

i	x_i	w_i
1	0.261799	-0.0844105
2	0.523599	-0.0819684

4. a)

i	x_i	w_{1i}
5	1.25	0.6431423
10	1.50	0.6832421
15	1.75	0.6922685

b)

i	x_i	w_{1i}
3	0.2356193	1.069066
6	0.4712387	1.079054
9	0.706858	1.029415

6. For Eq. (11.5), let $u_1(x) = y$ and $u_2(x) = y'$. Then

$$u_1'(x) = u_2(x), \quad a \le x \le b, \quad u_1(a) = \alpha$$

and

$$u_2'(x) = p(x)u_2(x) + q(x)u_1(x) + r(x), \quad a \le x \le b, \quad u_2(a) = 0.$$

For Eq. (11.6), let $v_1(x) = y$ and $v_2(x) = y'$. Then

$$v_1'(x) = v_2(x), \quad a \le x \le b, \quad v_1(a) = 0$$

and

$$v_2'(x) = p(x)v_2(x) + q(x)v_1(x), \quad a \le x \le b, \quad v_2(a) = 1.$$

Using the notation $u_{1,i} = u_1(x_i), u_{2,i} = u_2(x_i), v_{1,i} = v_1(x_i)$ and $v_{2,i} = v_2(x_i)$ leads to the equations in STEP 4 of Algorithm 11.1.

7. a) Let

$$f_1(x,y,z) = z, f_2(x,y,z) = p(x)z + q(x)y + r(x), \quad \text{and} \quad f_3(x,y,z) = p(x)z + q(x)y.$$

The steps are as follows:

STEP 1 Set

$h = (b-a)/N;$

$x_0 = a;$

$u_{1,0} = \alpha;$

$u_{2,0} = 0;$

$v_{1,0} = 0;$

$v_{2,0} = 1.$

STEP 2 For $i = 0$ to 2 set

$x_{i+1} = a + (i+1)h;$

$k_{1,1} = hf_1(x_i, u_{1,i}, u_{2,i});$

$k_{1,2} = hf_2(x_i, u_{1,i}, u_{2,i});$

$k_{2,1} = hf_1(x_i + h/2, u_{1,i} + k_{1,1}/2, u_{2,i} + k_{1,2}/2);$

$k_{2,2} = hf_2(x_i + h/2, u_{1,i} + k_{1,1}/2, u_{2,i} + k_{1,2}/2);$

$k_{3,1} = hf_1(x_i + h/2, u_{1,i} + k_{2,1}/2, u_{2,i} + k_{2,2}/2);$

$k_{3,2} = hf_2(x_i + h/2, u_{1,i} + k_{2,1}/2, u_{2,i} + k_{2,2}/2);$

$k_{4,1} = hf_1(x_{i+1}, u_{1,i} + k_{3,1}, u_{2,i} + k_{3,2});$

$k_{4,2} = hf_2(x_{i+1}, u_{1,i} + k_{3,1}, u_{2,i} + k_{3,2});$

$u_{1,i+1} = u_{1,i} + (k_{1,1} + 2k_{2,1} + 2k_{3,1} + k_{4,1})/6;$

$u_{2,i+1} = u_{2,i} + (k_{1,2} + 2k_{2,2} + 2k_{3,2} + k_{4,2})/6;$

$k'_{1,1} = hf_1(x_i, v_{1,i}, v_{2,i});$

$k'_{1,2} = hf_3(x_i, v_{1,i}, v_{2,i});$

$k'_{2,1} = hf_1(x_i + h/2, v_{1,i} + k'_{1,1}/2, v_{2,i} + k'_{1,2}/2);$

$k'_{2,2} = hf_3(x_i + h/2, v_{1,i} + k'_{1,1}/2, v_{2,i} + k'_{1,2}/2);$

$k'_{3,1} = hf_1(x_i + h/2, v_{1,i} + k'_{2,1}/2, v_{2,i} + k'_{2,2}/2);$

$k'_{3,2} = hf_3(x_i + h/2, v_{1,i} + k'_{2,1}/2, v_{2,i} + k'_{2,2}/2);$

$k'_{4,1} = hf_1(x_{i+1}, v_{1,i} + k'_{3,1}, v_{2,i} + k'_{3,2});$

$k'_{4,2} = hf_3(x_{i+1}, v_{1,i} + k'_{3,1}, v_{2,i} + k'_{3,2});$

$v_{1,i+1} = v_{1,i} + (k'_{1,1} + 2k'_{2,1} + 2k'_{3,1} + k'_{4,1})/6;$

$v_{2,i+1} = v_{2,i} + (k'_{1,2} + 2k'_{2,2} + 2k'_{3,2} + k'_{4,2})/6.$

STEP 3 For $i = 3, \ldots, N-1$ set

$\quad x_{i+1} = a + (i+1)h;$

$\quad UP = u_{1,i} + h[55f_1(x_i, u_{1,i}, u_{2,i}) - 59f_1(x_{i-1}, u_{1,i-1}, u_{2,i-1})$

$\qquad\qquad + 37f_1(x_{i-2}, u_{1,i-2}, u_{2,i-2}) - 9f_1(x_{i-3}, u_{1,i-3}, u_{2,i-3})]/24;$

$\quad UC = u_{2,i} + h[55f_2(x_i, u_{1,i}, u_{2,i}) - 59f_2(x_{i-1}, u_{1,i-1}, u_{2,i-1})$

$\qquad\qquad + 37f_2(x_{i-2}, u_{1,i-2}, u_{2,i-2}) - 9f_2(x_{i-3}, u_{1,i-3}, u_{2,i-3})]/24;$

$\quad u_{1,i+1} = u_{1,i} + h[9f_1(x_{i+1}, UP, UC) + 19f_1(x_i, u_{1,i}, u_{2,i}) - 5f_1(x_{i-1}, u_{1,i-1}, u_{2,i-1})$

$\qquad\qquad + f_1(x_{i-2}, u_{1,i-2}, u_{2,i-2})]/24;$

$\quad u_{2,i+1} = u_{2,i} + h[9f_2(x_{i+1}, UP, UC) + 19f_2(x_i, u_{1,i}, u_{2,i}) - 5f_2(x_{i-1}, u_{1,i-1}, u_{2,i-1})$

$\qquad\qquad + f_2(x_{i-2}, u_{1,i-2}, u_{2,i-2})]/24;$

$\quad VP = v_{1,i} + h[55f_1(x_i, v_{1,i}, v_{2,i}) - 59f_1(x_{i-1}, v_{1,i-1}, v_{2,i-1})$

$\qquad\qquad + 37f_1(x_{i-2}, v_{1,i-2}, v_{2,i-2}) - 9f_1(x_{i-3}, v_{1,i-3}, v_{2,i-3})]/24;$

$\quad VC = v_{2,i} + h[55f_3(x_i, v_{1,i}, v_{2,i}) - 59f_3(x_{i-1}, v_{1,i-1}, v_{2,i-1}) + 37f_3(x_{i-2}, v_{1,i-2}, v_{2,i-2})$

$\qquad\qquad - 9f_3(x_{i-3}, v_{1,i-3}, v_{2,i-3})]/24;$

$\quad v_{1,i+1} = v_{1,i} + h[9f_1(x_{i+1}, VP, VC) + 19f_1(x_i, v_{1,i}, v_{2,i}) - 5f_1(x_{i-1}, v_{1,i-1}, v_{2,i-1})$

$\qquad\qquad + f_1(x_{i-2}, v_{1,i-2}, v_{2,i-2})]/24;$

$\quad v_{2,i+1} = v_{2,i} + h[9f_3(x_{i+1}, VP, VC) + 19f_3(x_i, v_{1,i}, v_{2,i}) - 5f_3(x_{i-1}, v_{1,i-1}, v_{2,i-1})$

$\qquad\qquad + f_3(x_{i-2}, v_{1,i-2}, v_{2,i-2})]/24;$

STEP 4 Set

$\quad w_{1,0} = \alpha;$

$\quad w_{2,0} = (\beta - u_{1,N})/v_{1,N};$

$\quad \text{OUTPUT}(a, w_{1,0}, w_{2,0}).$

STEP 5 For $i = 1, \ldots, N$ set

$\quad w_{1,i} = u_{1,i} + w_{2,0}v_{1,i};$

$\quad w_{2,i} = u_{2,i} + w_{2,0}v_{2,i};$

$\quad \text{OUTPUT}(x_i, w_{1,i}, w_{2,i}).$

STEP 6 STOP

b) Let

$$f_1(x,y,z) = z, \quad f_2(x,y,z) = p(x)z + q(x)y + r(x), \quad \text{and} \quad f_3(x,y,z) = p(x)z + q(x)y.$$

The steps are as follows:

STEP 1 Set

$\quad h = hmax;$

$\quad t = a;$

$\quad u_{1,0} = \alpha;$

$\quad u_{2,0} = 0;$

$\quad v_{1,0} = 0;$

$v_{2,0} = 1;$

$i = 1;$

$flag = 1.$

STEP 2 While $(flag = 1)$ do STEPS 3–9

STEP 3 Set

$$p = u_{1,i-1};$$

$$q = u_{2,i-1};$$

$$k_{1,1} = hf_1(t, p, q);$$

$$k_{1,2} = hf_2(t, p, q);$$

$$k_{2,1} = hf_1(t + h/4, p + k_{1,1}/4, q + k_{1,2}/4);$$

$$k_{2,2} = hf_2(t + h/4, p + k_{1,1}/4, q + k_{1,2}/4);$$

$$k_{3,1} = hf_1(t + 3h/8, p + 3k_{1,1}/32 + 9k_{2,1}/32, q + 3k_{1,2}/32 + 9k_{2,2}/32);$$

$$k_{3,2} = hf_2(t + 3h/8, p + 3k_{1,1}/32 + 9k_{2,1}/32, q + 3k_{1,2}/32 + 9k_{2,2}/32);$$

$$k_{4,1} = hf_1(t + 12h/13, p + 1932k_{1,1}/2197 - 7200k_{2,1}/2197 + 7296k_{3,1}/2197,$$
$$q + 1932k_{1,2}/2197 - 7200k_{2,2}/2197 + 7296k_{3,2}/2197);$$

$$k_{4,2} = hf_2(t + 12h/13, p + 1932k_{1,1}/2197 - 7200k_{2,1}/2197 + 7296k_{3,1}/2197,$$
$$q + 1932k_{1,2}/2197 - 7200k_{2,2}/2197 + 7296k_{3,2}/2197);$$

$$k_{5,1} = hf_1(t + h, p + 439k_{1,1}/216 - 8k_{2,1} + 3680k_{3,1}/513 - 845k_{4,1}/4104,$$
$$q + 439k_{1,2}/216 - 8k_{2,2} + 3680k_{3,2}/513 - 845k_{4,2}/4104);$$

$$k_{5,2} = hf_2(t + h, p + 439k_{1,1}/216 - 8k_{2,1} + 3680k_{3,1}/513 - 845k_{4,1}/4104,$$
$$q + 439k_{1,2}/216 - 8k_{2,2} + 3680k_{3,2}/513 - 845k_{4,2}/4104);$$

$$k_{6,1} = hf_1(t + h/2, p - 8k_{1,1}/27 + 2k_{2,1} - 3544k_{3,1}/2565 + k_{4,1}/4$$
$$-11k_{5,1}/40, q - 8k_{1,2}/27 + 2k_{2,2} - 3544k_{3,2}/2565 + k_{4,2}/4 - 11k_{5,2}/40);$$

$$k_{6,2} = hf_2(t + h/2, p - 8k_{1,1}/27 + 2k_{2,1} - 3544k_{3,1}/2565 + k_{4,1}/4$$
$$-11k_{5,1}/40, q - 8k_{1,2}/27 + 2k_{2,2} - 3544k_{3,2}/2565 + k_{4,2}/4 - 11k_{5,2}/40);$$

STEP 4 Set

$$R_1 = |k_{1,1}/360 - 128k_{3,1}/4275 - 2197k_{4,1}/75240 + k_{5,1}/50 + 2k_{6,1}/55|/h;$$

$$R_2 = |k_{1,2}/360 - 128k_{3,2}/4275 - 2197k_{4,2}/75240 + k_{5,2}/50 + 2k_{6,2}/55|/h;$$

$$R = \max(R_1, R_2).$$

STEP 5 Set

$$p = v_{1,i-1};$$

$$q = v_{2,i-1};$$

Compute $k'_{1,1}, \ldots, k'_{6,2}$ as in STEP 3 except use f_3 in place of f_2.

STEP 6 Compute R'_1, R'_2, R' as in STEP 4 using $k'_{i,j}$ in place of $k_{i,j}$.

STEP 7 Set $R'' = \max (R, R')$.

STEP 8 Set $\delta = 0.84(TOL/R'')^{1/4}$.

STEP 9 If $(R'' \leq TOL)$ then set

$\quad\quad t = t + h;$

$\quad\quad u_{1,i} = u_{1,i-1} + 25k_{1,1}/216 + 1408k_{3,1}/2565 + 2197k_{4,1}/4104 - k_{5,1}/5;$

$\quad\quad u_{2,i} = u_{2,i-1} + 25k_{1,2}/216 + 1408k_{3,2}/2565 + 2197k_{4,2}/4104 - k_{5,2}/5;$

$\quad\quad v_{1,i} = v_{1,i-1} + 25k'_{1,1}/216 + 1408k'_{3,1}/2565 + 2197k'_{4,1}/4104 - k'_{5,1}/5;$

$\quad\quad v_{2,i} = v_{2,i-1} + 25k'_{1,2}/216 + 1408k'_{3,2}/2565 + 2197k'_{4,2}/4104 - k'_{5,2}/5;$

$\quad\quad x_i = t;$

$\quad\quad i = i + 1$

$\quad$ else

$\quad\quad\quad$ if $\delta \leq 0.1$ then set $h = 0.1h$

$\quad\quad\quad$ else

$\quad\quad\quad\quad$ if $\delta \geq 4$ then set $h = 4h$

$\quad\quad\quad\quad$ else set $h = \delta h$

$\quad\quad\quad$ if $h > hmax$ then set $h = hmax$

$\quad\quad\quad$ if $t \geq b$ then set $flag = 0$

$\quad\quad\quad$ else

$\quad\quad\quad\quad$ if $t + h > b$ then set $h = b - t$

$\quad\quad\quad$ else

$\quad\quad\quad\quad$ if $h < hmin$ then

$\quad\quad\quad\quad\quad$ OUTPUT('Minimal h exceeded.');

$\quad\quad\quad\quad\quad$ STOP.

STEP 10 Set

$\quad\quad N = i - 1;$

$\quad\quad w_{1,0} = \alpha;$

$\quad\quad w_{2,0} = (\beta - u_{1,N})/v_{1,N};$

$\quad\quad$ OUTPUT$(a, w_{1,0}, w_{2,0})$.

STEP 11 For $i = 1, \dots, N$ set

$\quad\quad w_{1,i} = u_{1,i} + w_{2,0}v_{1,i};$

$\quad\quad w_{2,i} = u_{2,i} + w_{2,0}v_{2,i};$

$\quad\quad$ OUTPUT$(x_i, w_{1,i}, w_{2,i})$.

STEP 12 STOP.

8. Using the Runge-Kutta-Fehlberg method with $TOL = 10^{-4}$ gives the results in the following tables.

(3a)

i	x_i	$w_{1,i}$	$w_{2,i}$
3	1.226032	0.1263455	1.791425
6	1.523631	0.4765760	0.7541501
8	1.723631	0.5958119	0.4672616
11	2.000000	0.6715967	0.3046429

(3b)

i	x_i	$w_{1,i}$	$w_{2,i}$
11	1.021761	0.007188846	-0.03471082
17	2.196027	2.406999×10^{-5}	-1.162202×10^{-4}
22	3.196027	1.873674×10^{-7}	-9.047272×10^{-7}
27	4.196027	1.443500×10^{-9}	-7.055366×10^{-9}

(3c)

i	x_i	$w_{1,i}$	$w_{2,i}$
4	0.3311956	0.7320051	-1.486103
7	0.6496132	0.6223601	0.5085259
9	0.9147805	0.8764586	1.344295

(3d)

i	x_i	$w_{1,i}$	$w_{2,i}$
3	0.3	-0.04603469	-0.02847177
6	0.6	-0.03460934	0.08130043
9	0.9	-0.007869820	0.08447164

(4a)

i	x_i	$w_{1,i}$	$w_{2,i}$
3	1.289835	0.6535320	0.2350112
5	1.489835	0.6824695	0.07869523
8	1.789835	0.6926519	0.007697520

(4b)

i	x_i	$w_{1,i}$	$w_{2,i}$
2	0.2000000	1.062358	0.2072875
5	0.5000000	1.076167	-0.1159190
8	0.7853982	1.000000	-0.4142136

Using the Adams-Bashforth and Adams-Moulton Predictor-Corrector method:

(3a)

i	x_i	$w_{1,i}$	$w_{2,i}$
5	1.25	0.1677087	1.655720
10	1.50	0.4582671	0.8014088
15	1.75	0.6078048	0.4404206
20	2.00	0.6931472	0.2609362

(3b)

i	x_i	$w_{1,i}$	$w_{2,i}$
5	1.0	−0.01261295	0.06088475
10	2.0	0.00206541	−0.0100092
15	3.0	−0.000108012	0.000437854
20	4.0	−0.000226841	0.000903692

(3c)

i	x_i	$w_{1,i}$	$w_{2,i}$
3	0.3	0.7834958	−1.800364
6	0.6	0.6019723	0.2993897
9	0.9	0.8567850	1.307813

(3d)

i	x_i	$w_{1,i}$	$w_{2,i}$
3	0.3	−0.04603212	−0.02847704
6	0.6	−0.03460412	0.08128354
9	0.9	−0.007868194	0.08445786

(4a)

i	x_i	$w_{1,i}$	$w_{2,i}$
5	1.25	0.6431631	0.2879506
10	1.50	0.6832616	0.07400812
15	1.75	0.6922771	0.01161611

(4b)

i	x_i	$w_{1,i}$	$w_{2,i}$
3	0.2356194	1.069066	0.1693234
6	0.4712389	1.079056	−0.08492378
9	0.7068583	1.029416	−0.3344781

10. Since $y_2(a) = 0$ and $y_2(b) = 0$, the boundary value problem

$$y'' = p(x)y' + q(x)y, \quad a \le x \le b \quad, \quad y(a) = 0, \quad y(b) = 0$$

has $y = 0$ as a unique solution, so $y_2 \equiv 0$.

11. Solve the three initial value problems:

$$y_1'' = p(x)y_1' + q(x)y_1 + r(x), \quad a \le x \le b, \quad y_1(a) = 0, \quad y_1'(a) = 0;$$

$$y_2'' = p(x)y_2' + q(x)y_2, \quad a \le x \le b, \quad y_2(a) = 1, \quad y_2'(a) = 0;$$

and

$$y_3'' = p(x)y_3' + q(x)y_3, \quad a \le x \le b, \quad y_3(a) = 0, \quad y_3'(a) = 1.$$

Let $y(x) = y_1(x) + k_1 y_2(x) + k_2 y_3(x)$. Clearly, y satisfies the differential equation. For the boundary conditions, we need

$$\alpha = \alpha_1 y(a) + \beta_1 y'(a) = \alpha_1 k_1 + \beta_1 k_2$$

and

$$\beta = \alpha_2 y(b) + \beta_2 y'(b)$$
$$= \alpha_2 y_1(b) + \alpha_2 k_1 y_2(b) + \alpha_2 k_2 y_3(b) + \beta_2 y_1'(b) + \beta_2 k_1 y_2'(b) + \beta_2 k_2 y_3'(b).$$

Thus, the linear system

$$\begin{bmatrix} \alpha_1 & \beta_1 \\ \alpha_2 y_2(b) + \beta_2 y_2'(b) & \alpha_2 y_3(b) + \beta_2 y_3'(b) \end{bmatrix} \begin{bmatrix} k_1 \\ k_2 \end{bmatrix} = \begin{bmatrix} \alpha \\ \beta - \alpha_2 y_1(b) - \beta_2 y_1'(b) \end{bmatrix}$$

must be solved for k_1 and k_2.

12. a) $u(3) \simeq 36.66668,$ b) $u(3) \simeq 36.66667,$ c) $u(3) = 36.\overline{6}.$

Exercise Set 11.2 (PAGE 573)

2.

x_i	w_{1i}	w_{2i}
1.2	−0.981211	0.182305
1.4	−0.928936	0.336455

4. Modify Algorithm 11.2 as follows:

STEP 1 Set $h = (b-a)/N$;

$\qquad k = 2$;

$\qquad TK1 = (\beta - \alpha)/(b-a)$.

STEP 2 Set $w_{1,0} = \alpha$;

$\qquad\qquad w_{2,0} = TK1$.

STEP 3 For $i = 1, \ldots, N$ do STEPS 4 and 5.

$\qquad$ STEP 4 Set $x = a + (i-1)h$.

$\qquad$ STEP 5 Set

$\qquad\qquad k_{1,1} = hw_{2,i-1}$;

$\qquad\qquad k_{1,2} = hf(x, w_{1,i-1}, w_{2,i-1})$;

$\qquad\qquad k_{2,1} = h(w_{2,i-1} + k_{1,2}/2)$;

$\qquad\qquad k_{2,2} = hf(x + h/2, w_{1,i-1} + k_{1,1}/2, w_{2,i-1} + k_{1,2}/2)$;

$\qquad\qquad k_{3,1} = h(w_{2,i-1} + k_{2,2}/2)$;

$\qquad\qquad k_{3,2} = hf(x + h/2, w_{1,i-1} + k_{2,1}/2, w_{2,i-1} + k_{2,2}/2)$;

$\qquad\qquad k_{4,1} = h(w_{2,i-1} + k_{3,2}/2)$;

$\qquad\qquad k_{4,2} = hf(x + h/2, w_{1,i-1} + k_{3,1}, w_{2,i-1} + k_{3,2})$;

$\qquad\qquad w_{1,i} = w_{1,i-1} + (k_{1,1} + 2k_{2,1} + 2k_{3,1} + k_{4,1})/6$;

$\qquad\qquad w_{2,i} = w_{2,i-1} + (k_{1,2} + 2k_{2,2} + 2k_{3,2} + k_{4,2})/6$.

STEP 6 Set $TK2 = TK1 + (\beta - w_{1,N})/(b-a)$.

STEP 7 While $(k \leq M)$ do STEPS 8–15.

$\qquad$ STEP 8 Set $w_{2,0} = TK2$;

$\qquad\qquad HOLD = w_{1,N}$.

$\qquad$ STEP 9 For $i = 1, \ldots, N$ do STEPS 10 and 11.

$\qquad\qquad$ STEP 10 (Same as STEP 4)

$\qquad\qquad$ STEP 11 (Same as STEP 5)

$\qquad$ STEP 12 If $|w_{1,N} - \beta| \leq TOL$ then do STEPS 13 and 14.

$\qquad\qquad$ STEP 13 For $i = 0, \ldots, N$ set

$\qquad\qquad\qquad x = a + ih$;

$\qquad\qquad\qquad$ OUTPUT$(x, w_{1,i}, w_{2,i})$.

$\qquad\qquad$ STEP 14 STOP.

STEP 15 Set
$$TK = TK2 - (w_{1,N} - \beta)(TK2 - TK1)/(w_{1.N} - HOLD);$$
$$TK1 = TK2;$$
$$TK2 = TK;$$
$$k = k + 1.$$

STEP 16 OUTPUT('Maximum number of iterations exceeded.');
    STOP.

5. 3b)

x_i	w_{1i}	w_{2i}
2.0	0.2000039	−0.03999572
3.0	0.1666753	−0.02777229
4.0	0.1428719	−0.02040104
5.0	0.1250227	−0.05161591

3d)

x_i	w_{1i}	w_{2i}
1.25	2.049997	0.3599871
1.50	2.166655	0.5555039
1.75	1.321388	0.6732749
2.00	2.499851	0.7491976

6. a) Let

$$f_1(x,y,z) = z, f_2(x,y,z) = f(x,y,z), \quad \text{and} \quad f_3(x,y,z,p,q) = f_y(x,p,q)y + f_{y'}(x,p,q)z.$$

Perform the following steps:

STEP 1 Set
$$h = (b - a)/N;$$
$$k = 1;$$
$$TK = (\beta - \alpha)/(b - a).$$

STEP 2 While $(k \leq M)$ do STEPS 3–6.
    STEP 3 Set
$$u_{1,0} = \alpha;$$
$$u_{2,0} = TK;$$

$$v_{1,0} = 0;$$
$$v_{2,0} = 1;$$
$$x_0 = a.$$

STEP 4 For $i = 0$ to 2 set

$$x_{i+1} = a + (i+1)h;$$
$$k_{1,1} = hf_1(x_i, u_{1,i}, u_{2,i});$$
$$k_{1,2} = hf_2(x_i, u_{1,i}, u_{2,i});$$
$$k_{2,1} = hf_1(x_i + h/2, u_{1,i} + k_{1,1}/2, u_{2,i} + k_{1,2}/2);$$
$$k_{2,2} = hf_2(x_i + h/2, u_{1,i} + k_{1,1}/2, u_{2,i} + k_{1,2}/2);$$
$$k_{3,1} = hf_1(x_i + h/2, u_{1,i} + k_{2,1}/2, u_{2,i} + k_{2,2}/2);$$
$$k_{3,2} = hf_2(x_i + h/2, u_{1,i} + k_{2,1}/2, u_{2,i} + k_{2,2}/2);$$
$$k_{4,1} = hf_1(x_{i+1}, u_{1,i} + k_{3,1}, u_{2,i} + k_{3,2});$$
$$k_{4,2} = hf_2(x_{i+1}, u_{1,i} + k_{3,1}, u_{2,i} + k_{3,2});$$
$$u_{1,i+1} = u_{1,i} + (k_{1,1} + 2k_{2,1} + 2k_{3,1} + k_{4,1})/6;$$
$$u_{2,i+1} = u_{2,i} + (k_{1,2} + 2k_{2,2} + 2k_{3,2} + k_{4,2})/6;$$
$$k'_{1,1} = hf_1(x_i, v_{1,i}, v_{2,i});$$
$$k'_{1,2} = hf_3(x_i, v_{1,i}, v_{2,i}, u_{1,i}, u_{2,i});$$
$$k'_{2,1} = hf_1(x_i + h/2, v_{1,i} + k'_{1,1}/2, v_{2,i} + k'_{1,2}/2);$$
$$k'_{2,2} = hf_3(x_i + h/2, v_{1,i} + k'_{1,1}/2, v_{2,i} + k'_{1,2}/2, u_{1,i} + k_{1,1}/2, u_{2,i} + k_{1,2}/2);$$
$$k'_{3,1} = hf_1(x_i + h/2, v_{1,i} + k'_{2,1}/2, v_{2,i} + k'_{2,2}/2);$$
$$k'_{3,2} = hf_3(x_i + h/2, v_{1,i} + k'_{2,1}/2, v_{2,i} + k'_{2,2}/2, u_{1,i} + k_{2,1}/2, u_{2,i} + k_{2,2}/2);$$
$$k'_{4,1} = hf_1(x_{i+1}, v_{1,i} + k'_{3,1}, v_{2,i} + k'_{3,2});$$
$$k'_{4,2} = hf_3(x_{i+1}, v_{1,i} + k'_{3,1}, v_{2,i} + k'_{3,2}, u_{1,i} + k_{3,1}, u_{2,i} + k_{3,2});$$
$$v_{1,i+1} = v_{1,i} + (k'_{1,1} + 2k'_{2,1} + 2k'_{3,1} + k'_{4,1})/6;$$
$$v_{2,i+1} = v_{2,i} + (k'_{1,2} + 2k'_{2,2} + 2k'_{3,2} + k'_{4,2})/6.$$

STEP 5 For $i = 3, \ldots, N-1$

$$x_{i+1} = a + (i+1)h;$$
$$UP = u_{1,i} + h[55f_1(x_i, u_{1,i}, u_{2,i}) - 59f_1(x_{i-1}, u_{1,i-1}, u_{2,i-1})$$
$$+ 37f_1(x_{i-2}, u_{1,i-2}, u_{2,i-2}) - 9f_1(x_{i-3}, u_{1,i-3}, u_{2,i-3})]/24;$$
$$UC = u_{2,i} + h[55f_2(x_i, u_{1,i}, u_{2,i}) - 59f_2(x_{i-1}, u_{1,i-1}, u_{2,i-1})$$
$$+ 37f_2(x_{i-2}, u_{1,i-2}, u_{2,i-2}) - 9f_2(x_{i-3}, u_{1,i-3}, u_{2,i-3})]/24;$$
$$u_{1,i+1} = u_{1,i} + h[9f_1(x_{i+1}, UP, UC) + 19f_1(x_i, u_{1,i}, u_{2,i})$$
$$- 5f_1(x_{i-1}, u_{1,i-1}, u_{2,i-1}) + f_1(x_{i-2}, u_{1,i-2}, u_{2,i-2})]/24;$$
$$u_{2,i+1} = u_{2,i} + h[9f_2(x_{i+1}, UP, UC) + 19f_2(x_i, u_{1,i}, u_{2,i})$$
$$- 5f_2(x_{i-1}, u_{1,i-1}, u_{2,i-1}) + f_2(x_{i-2}, u_{1,i-2}, u_{2,i-2})]/24;$$
$$VP = v_{1,i} + h[55f_1(x_i, v_{1,i}, v_{2,i}) - 59f_1(x_{i-1}, v_{1,i-1}, v_{2,i-1})$$
$$+ 37f_1(x_{i-2}, v_{1,i-2}, v_{2,i-2}) - 9f_1(x_{i-3}, v_{1,i-3}, v_{2,i-3})]/24;$$
$$VC = v_{2,i} + h[55f_3(x_i, v_{1,i}, v_{2,i}, u_{1,i}, u_{2,i})$$

$$-59f_3(x_{i-1}, v_{1,i-1}v_{2,i-1}, u_{1,i-1}, u_{2,i-1}),$$
$$+37f_3(x_{i-2}, v_{1,i-2}, v_{2,i-2}, u_{1,i-2}, u_{2,i-2})$$
$$-9f_3(x_{i-3}, v_{1,i-3}, v_{2,i-3}, u_{1,i-3}, u_{2,i-3})]/24;$$
$$v_{1,i+1} = v_{1,i} + h[9f_1(x_{i+1}, VP, VC) + 19f_1(x_i, v_{1,i}, v_{2,i})$$
$$-5f_1(x_{i-1}, v_{1,i-1}, v_{2,i-1}) + f_1(x_{i-2}, v_{1,i-2}, v_{2,i-2})]/24;$$
$$v_{2,i+1} = v_{2,i} + h[9f_3(x_{i+1}, VP, VC, u_{1,i+1}, u_{2,i+1})$$
$$19f_3(x_i, v_{1,i}, v_{2,i}, u_{1,i}, u_{2,i})$$
$$-5f_3(x_{i-1}, v_{1,i-1}, v_{2,i-1}, u_{1,i-1}, u_{2,i-1})$$
$$+f_3(x_{i-2}, v_{1,i-2}, v_{2,i-2}, u_{1,i-2}, u_{2,i-2})]/24;$$

STEP 6 If $|u_{1,N} - \beta| \leq TOL$ then for $i = 0, \ldots, N$

OUTPUT$(x_i, u_{1,i}, u_{2,i})$;

STOP

else

set $TK = TK - (u_{1,N} - \beta)/v_{1,N}$;

$k = k + 1$.

STEP 7 OUTPUT('Maximum number of iterations exceeded.');

STOP.

b) Let

$$f_1(x, y, z) = z, \quad f_2(x, y, z) = f(x, y, z), \quad \text{and} \quad f_3(x, y, z, p, q) = f_y(x, p, q)y + f_{y'}(x, p, q)z.$$

Perform the following steps:

STEP 1 Set

$h = hmax$;

$k = 1$;

$TK = (\beta - \alpha)/(b - a)$;

$u_{1,0} = \alpha$;

$v_{1,0} = 0$;

$v_{2,0} = 1$.

STEP 2 While $(k \leq M)$ do STEPS 3–16.

STEP 3 Set

$t = a$;

$i = 1$;

$u_{2,0} = TK$;

$flag = 1$.

STEP 4 While $(flag = 1)$ do STEPS 5–11

STEP 5 Set

120

$$p = u_{1,i-1};$$

$$q = u_{2,i-1};$$

$$k_{1,1} = hf_1(t, p, q);$$

$$k_{1,2} = hf_2(t, p, q);$$

$$k_{2,1} = hf_1(t + h/4, p + k_{1,1}/4, q + k_{1,2}/4);$$

$$k_{2,2} = hf_2(t + h/4, p + k_{1,1}/4, q + k_{1,2}/4);$$

$$k_{3,1} = hf_1(t + 3h/8, p + 3k_{1,1}/32 + 9k_{2,1}/32, q + 3k_{1,2}/32 + 9k_{2,2}/32);$$

$$k_{3,2} = hf_2(t + 3h/8, p + 3k_{1,1}/32 + 9k_{2,1}/32, q + 3k_{1,2}/32 + 9k_{2,2}/32);$$

$$k_{4,1} = hf_1(t + 12h/13, p + 1932k_{1,1}/2197 - 7200k_{2,1}/2197 + 7296k_{3,1}/2197,$$
$$q + 1932k_{1,2}/2197 - 7200k_{2,2}/2197 + 7296k_{3,2}/2197);$$

$$k_{4,2} = hf_2(t + 12h/13, p + 1932k_{1,1}/2197 - 7200k_{2,1}/2197 + 7296k_{3,1}/2197,$$
$$q + 1932k_{1,2}/2197 - 7200k_{2,2}/2197 + 7296k_{3,2}/2197);$$

$$k_{5,1} = hf_1(t + h, p + 439k_{1,1}/216 - 8k_{2,1} + 3680k_{3,1}/513 - 845k_{4,1}/4104,$$
$$q + 439k_{1,2}/216 - 8k_{2,2} + 3680k_{3,2}/513 - 845k_{4,2}/4104);$$

$$k_{5,2} = hf_2(t + h, p + 439k_{1,1}/216 - 8k_{2,1} + 3680k_{3,1}/513 - 845k_{4,1}/4104,$$
$$q + 439k_{1,2}/216 - 8k_{2,2} + 3680k_{3,2}/513 - 845k_{4,2}/4104);$$

$$k_{6,1} = hf_1(t + h/2, p - 8k_{1,1}/27 + 2k_{2,1} - 3544k_{3,1}/2565 + k_{4,1}/4$$
$$-11k_{5,1}/40, q - 8k_{1,2}/27 + 2k_{2,2} - 3544k_{3,2}/2565 + k_{4,2}/4 - 11k_{5,2}/40);$$

$$k_{6,2} = hf_2(t + h/2, p - 8k_{1,1}/27 + 2k_{2,1} - 3544k_{3,1}/2565 + k_{4,1}/4$$
$$-11k_{5,1}/40, q - 8k_{1,2}/27 + 2k_{2,2} - 3544k_{3,2}/2565 + k_{4,2}/4 - 11k_{5,2}/40);$$

STEP 6 Set

$$R_1 = |k_{1,1}/360 - 128k_{3,1}/4275 - 2197k_{4,1}/75240 + k_{5,1}/50 + 2k_{6,1}/55|/h;$$

$$R_2 = |k_{1,2}/360 - 128k_{3,2}/4275 - 2197k_{4,2}/75240 + k_{5,2}/50 + 2k_{6,2}/55|/h;$$

$$R = \max(R_1, R_2).$$

STEP 7 Set

$$p' = v_{1,i-1};$$

$$q' = v_{2,i-1};$$

$$k'_{1,1} = hf_1(t, p', q');$$

$$k'_{1,2} = hf_3(t, p', q', p, q);$$

$$k'_{2,1} = hf_1(t + h/4, p' + k'_{1,1}/4, q' + k'_{1,2}/4);$$

$$k'_{2,2} = hf_3(t + h/4, p' + k'_{1,1}/4, q' + k'_{1,2}/4, p + k_{1,1}/4, q + k_{1,2}/4);$$

$$k'_{3,1} = hf_1(t + 3h/8, p' + 3k'_{1,1}/32 + 9k'_{2,1}/32, q' + 3k'_{1,2}/32 + 9k'_{2,2}/32);$$

$$k'_{3,2} = hf_3(t + 3h/8, p' + 3k'_{1,1}/32 + 9k'_{2,1}/32, q' + 3k'_{1,2}/32$$
$$+9k'_{2,2}/32, p + 3k_{1,1}/32 + 9k_{2,1}/32, q + 3k_{1,2}/32 + 9k_{2,2}/32);$$

$$k'_{4,1} = hf_1(t + 12h/13, p' + 1932k'_{1,1}/2197 - 7200k'_{2,1}/2197$$
$$+7296k'_{3,1}/2197, q' + 1932k'_{1,2}/2197 - 7200k'_{2,2}/2197 + 7296k'_{3,2}/2197);$$

$$k'_{4,2} = hf_3(t + 12h/13, p' + 1932k'_{1,1}/2197 - 7200k'_{2,1}/2197$$

$$+7296k'_{3,1}/2197, q' + 1932k'_{1,2}/2197 - 7200k'_{2,2}/2197$$
$$+7296k'_{3,2}/2197, p + 1932k_{1,1}/2197 - 7200k_{2,1}/2197$$
$$+7296k_{3,1}/2197, q + 1932k_{1,2}/2197$$
$$-7200k_{2,2}/2197 + 7296k_{3,2}/2197);$$

$$k'_{5,1} = hf_1(t + h, p' + 439k'_{1,1}/216 - 8k'_{2,1} + 3680k'_{3,1}/513 - 845k'_{4,1}/4104,$$
$$q' + 439k'_{1,2}/216 - 8k'_{2,2} + 3680k'_{3,2}/513 - 845k'_{4,2}/4104);$$

$$k'_{5,2} = hf_3(t + h, p' + 439k'_{1,1}/216 - 8k'_{2,1} + 3680k'_{3,1}/513 - 845k'_{4,1}/4104,$$
$$q' + 439k'_{1,2}/216 - 8k'_{2,2} + 3680k'_{3,2}/513 - 845k'_{4,2}/4104,$$
$$p + 439k_{1,1}/216 - 8k_{2,1} + 3680k_{3,1}/513 - 845k_{4,1}/4104,$$
$$q + 439k_{1,2}/216 - 8k_{2,2} + 3680k_{3,2}/513 - 845k_{4,2}/4104);$$

$$k'_{6,1} = hf_1(t + h/2, p' - 8k'_{1,1}/27 + 2k'_{2,1} - 3544k'_{3,1}/2565 + k'_{4,1}/4$$
$$-11k'_{5,1}/40, q' - 8k'_{1,2}/27 + 2k'_{2,2} - 3544k'_{3,2}/2565 + k'_{4,2}/4 - 11k'_{5,2}/40);$$

$$k'_{6,2} = hf_3(t + h/2, p' - 8k'_{1,1}/27 + 2k'_{2,1} - 3544k'_{3,1}/2565 + k'_{4,1}/4$$
$$-11k'_{5,1}/40, q' - 8k'_{1,2}/27 + 2k'_{2,2} - 3544k'_{3,2}/2565 + k'_{4,2}/4$$
$$-11k'_{5,2}/40, p - 8k_{1,1}/27 + 2k_{2,1} - 3544k_{3,1}/2565$$
$$+k_{4,1}/4 - 11k_{5,1}/40, q - 8k_{1,2}/27 + 2k_{2,2} - 3544k_{3,2}/2565$$
$$+k_{4,2}/4 - 11k_{5,2}/40);$$

STEP 8 Compute R'_1, R'_2, R' as in STEP 6 using $k'_{i,j}$ in place of $k_{i,j}$.

STEP 9 Set $R'' = \max(R, R')$.

STEP 10 Set $\delta = 0.84(TOL/R'')^{1/4}$.

STEP 11 If $(R'' \leq TOL)$ then set

$t = t + h$;

$u_{1,i} = u_{1,i-1} + 25k_{1,1}/216 + 1408k_{3,1}/2565 + 2197k_{4,1}/4104 - k_{5,1}/5$;

$u_{2,i} = u_{2,i-1} + 25k_{1,2}/216 + 1408k_{3,2}/2565 + 2197k_{4,2}/4104 - k_{5,2}/5$;

$v_{1,i} = v_{1,i-1} + 25k'_{1,1}/216 + 1408k'_{3,1}/2565 + 2197k'_{4,1}/4104 - k'_{5,1}/5$;

$v_{2,i} = v_{2,i-1} + 25k'_{1,2}/216 + 1408k'_{3,2}/2565 + 2197k'_{4,2}/4104 - k'_{5,2}/5$;

$x_i = t$;

$i = i + 1$

   else

      if $\delta \leq 0.1$ then set $h = 0.1h$

      else

         if $\delta \geq 4$ then set $h = 4h$

         else set $h = \delta h$

      if $h > hmax$ then set $h = hmax$

      if $t \geq b$ then set $flag = 0$

      else

$$\text{if } t + h > b \text{ then set } h = b - t$$

else

if $h < hmin$ then

OUTPUT('Minimal h exceeded.');

STOP.

STEP 12 Set $N = i - 1$.

STEP 13 If $|u_{1,N} - \beta| \le TOL$ then do STEPS 14 and 15.

STEP 14 For $i = 0, \ldots, N$

OUTPUT($x_i, u_{1,i}, u_{2,i}$).

STEP 15 STOP.

STEP 16 Set $TK = TK - (u_{1,N} - \beta)/v_{1,N}$;

$k = k + 1$.

STEP 17 OUTPUT('Maximum number of iterations exceeded.');

STOP.

7. Using the Runge-Kutta-Fehlberg method with $TOL = 10^{-4}$ gives the entries in the following tables:

(3a)

i	x_i	$w_{1,i}$	$w_{2,i}$
2	1.2	−0.7142853	−0.2550998
4	1.4	−0.7692299	−0.2958556
6	1.6	−0.8333319	−0.3472196
8	1.8	−0.9090890	−0.4132202

(3b)

i	x_i	$w_{1,i}$	$w_{2,i}$
8	2.6	0.1785798	-0.03188158
16	4.2	0.1389095	-0.01928072
20	5.0	0.1250289	-0.01561356

(3c)

i	x_i	$w_{1,i}$	$w_{2,i}$
2	1.2	0.4545454	−0.2066116
5	1.5	0.4000000	−0.1600000
7	1.7	0.3703703	−0.1371742

i	x_i	$w_{1,i}$	$w_{2,i}$
4	1.269024	2.057030	0.3790534
10	1.602657	2.226621	0.6106757
18	1.860090	2.397701	0.7109931
25	2.000000	2.500006	0.7500341

Using the Adams-Bashforth and Adams-Moulton Predictor-Corrector method gives:

(3a)

i	x_i	$w_{1,i}$	$w_{2,i}$
5	1.25	-0.7272727	-0.2644629
10	1.50	-0.8000000	-0.3200001
15	1.75	-0.8888890	-0.3950620

(3b)

i	x_i	$w_{1,i}$	$w_{2,i}$
10	2.6	0.1785717	-0.03188714
20	4.2	0.1388903	-0.01928926
25	5.0	0.1250021	-0.01562399

(3c)

i	x_i	$w_{1,i}$	$w_{2,i}$
2	1.2	0.4545458	-0.2066105
5	1.5	0.3999997	-0.1599976
7	1.7	0.3703699	-0.1371712

(3d)

i	x_i	$w_{1,i}$	$w_{2,i}$
5	1.25	2.049999	0.3600068
10	1.50	2.166665	0.5555639
15	1.75	2.321428	0.6734744
20	2.00	2.500000	0.7500041

Exercise Set 11.3 (PAGE 579)

2.

a)

i	x_i	w_i
5	1.25	0.16797186
10	1.50	0.45842388
15	1.75	0.60787334

b)

x_i	w_i
1.0	6.332971×10^{-3}
2.0	4.010654×10^{-5}
3.0	2.539917×10^{-7}
4.0	1.604072×10^{-9}

c)

i	x_i	w_i
2	0.2	1.018096
5	0.5	0.5942743
7	0.7	0.6514520

d)

i	x_i	w_i
3	0.3	-0.04617547
6	0.6	-0.03471589
9	0.9	-0.007894152

4.

i	x_i	$w_i(h = 0.1)$
3	0.3	0.05572807
6	0.6	0.00310518
9	0.9	0.00016516

i	x_i	$w_i(h = 0.05)$
6	0.3	0.05132396
12	0.6	0.00263406
19	0.9	0.00013340

6. First,

$$\left|\frac{h}{2}p(x_i)\right| \le \frac{hL}{2} < 1$$

so

$$\left|-1-\frac{h}{2}p(x_i)\right| = 1+\frac{h}{2}p(x_i) \quad \text{and} \quad \left|-1+\frac{h}{2}p(x_i)\right| = 1-\frac{h}{2}p(x_i).$$

Therefore,

$$\left|-1-\frac{h}{2}p(x_i)\right| + \left|-1+\frac{h}{2}p(x_i)\right| = 2 \le 2 + h^2 q(x_i)$$

for $2 \le i \le N - 1$.

Since

$$\left| -1 + \frac{h}{2}p(x_i) \right| < 2 \le 2 + h^2 q(x_i) \quad \text{and} \quad \left| -1 - \frac{h}{2}p(x_N) \right| < 2 \le 2 + h^2 q(x_N),$$

Theorem 6.30 implies that the linear system (11.27) has a unique solution.

8.

x_i	w_i
10.0	0.1098549
20.0	0.1761424
25.0	0.1849608
30.0	0.1761424
40.0	0.1098549

Exercise Set 11.4 (PAGE 586)

2. a)

i	x_i	w_i
5	1.25	−0.7272810
10	1.50	−0.8000128
15	1.75	−0.8889005

b)

i	x_i	w_i
5	2.0	0.2000223
10	3.0	0.1666879
15	4.0	0.1428695
20	5.0	0.1250000

c)

i	x_i	w_i
2	1.2	0.4545563
5	1.5	0.4000130
7	1.7	0.3703798

d)

i	x_i	w_i
6	1.25	2.050045
11	1.50	2.166697
16	1.75	2.321441

4. The Jacobian matrix $J = (a_{i,j})$ is tridiagonal with entries given in (11.32). So

$$a_{1,1} = 2 + h^2 f_y(x_1, w_1, \frac{1}{2h}(w_2 - \alpha)),$$

$$a_{1,2} = -1 + \frac{h}{2} f_{y'}(x_1, w_1, \frac{1}{2h}(w_2 - \alpha)),$$

$$a_{i,i-1} = -1 - \frac{h}{2} f_{y'}(x_i, w_i, \frac{1}{2h}(w_{i+1} - w_{i-1})), \quad \text{for } 2 \le i \le N - 1$$

$$a_{i,i} = 2 + h^2 f_y(x_i, w_i, \frac{1}{2h}(w_{i+1} - w_{i-1})), \quad \text{for } 2 \le i \le N - 1$$

$$a_{i,i+1} = -1 + \frac{h}{2} f_{y'}(x_i, w_i, \frac{1}{2h}(w_{i+1} - w_{i-1})), \quad \text{for } 2 \le i \le N - 1$$

$$a_{N,N-1} = -1 + \frac{h}{2} f_{y'}(x_N, w_N, \frac{1}{2h}(\beta - w_{N-1})),$$

$$a_{N,N} = 2 + h^2 f_y(x_N, w_N, \frac{1}{2h}(\beta - w_{N-1})).$$

Thus, $|a_{i,i}| \ge 2 + h^2 \delta$ for $i = 1, \ldots, N$. Since $|f_{y'}(x, y, y')| \le L$ and $h < 2/L$,

$$|\frac{h}{2} f_{y'}(x, y, y')| \le \frac{hL}{2} < 1.$$

Thus,

$$|a_{1,2}| = \left| -1 + \frac{h}{2} f_{y'}(x_1, w_1, \frac{1}{2h}(w_2 - \alpha)) \right| < 2 < |a_{1,1}|,$$

$$|a_{i,i-1}| + |a_{i,i+1}| = -a_{i,i-1} - a_{i,i+1}$$
$$= 1 + \frac{h}{2} f_{y'}(x_i, w_i, \frac{1}{2h}(w_{i+1} - w_{i-1})) + 1 - \frac{h}{2} f_{y'}(x_i, w_i, \frac{1}{2h}(w_{i+1} - w_{i-1}))$$
$$= 2 \le |a_{i,i}|.$$

and

$$|a_{N,N-1}| = -a_{N,N-1} = 1 + \frac{h}{2} f_{y'}(x_N, w_N, \frac{1}{2h}(\beta - w_{N-1})) < 2 < |a_{N,N}|.$$

Hence by Theorem 6.30 J is nonsingular.

Exercise Set 11.5 (PAGE 601)

2. a)

i	c_i
0	-4.38127×10^{-3}
1	-5.29927×10^{-2}
2	-6.86142×10^{-2}
3	-5.02453×10^{-2}
4	-3.10118×10^{-3}

b)

i	c_i
0	-1.38889×10^{-2}
1	-0.138889
2	-0.180556
3	-0.138889
4	-1.38889×10^{-2}

4. a)

x_i	Approximation	x_i	Approximation
0.0	0.00000000	0.6	0.058259848
0.1	0.014766286	0.7	0.054507290
0.2	0.028679521	0.8	0.044294407
0.3	0.040878126	0.9	0.026518200
0.4	0.050483351	1.0	0.00000000
0.5	0.056590497		

b)

x_i	Approximation	x_i	Approximation	x_i	Approximation
0.00	0.00000000	0.35	-0.22750000	0.70	-0.20999999
0.05	-0.047500000	0.40	-0.24000000	0.75	-0.18749998
0.10	-0.090000000	0.45	-0.24749999	0.80	-0.15999998
0.15	-0.12750000	0.50	-0.24999999	0.85	-0.12749998
0.20	-0.16000000	0.55	-0.24749999	0.90	-0.089999975
0.25	-0.18750000	0.60	-0.23999999	0.95	-0.047499975
0.30	-0.21000000	0.65	-0.22749999	1.00	0.00000000

c)

i	c_i	$\phi(x_i)$
0	0.01996389	0.0
1	0.3600538	0.08564579
2	0.5953200	0.1450149
3	0.7390230	0.1814269
4	0.8028321	0.1978077
5	0.7970312	0.1967345
6	0.7306678	0.1804753
7	0.6116993	0.1510244
8	0.4471153	0.1101343
9	0.2430548	0.05934321
10	0.004894037	0.0

d)

i	$\phi(x_i)$	i	$\phi(x_i)$
0	0	11	−0.5417659
1	−0.08335051	12	−0.5376978
2	−0.1613111	13	−0.5217817
3	−0.2334671	14	−0.4931704
4	−0.2993758	15	−0.4509614
5	−0.3585641	16	−0.3941928
6	−0.4105269	17	−0.3218398
7	−0.4547250	18	−0.2328110
8	−0.4905828	19	−0.1259437
9	−0.5174863	20	0
10	−0.5347803		

6. If $\sum_{i=1}^{n} c_i \phi_i(x) = 0$ for $0 \le x \le 1$, then for any j we have $\sum_{i=1}^{n} c_i \phi_i(x_j) = 0$.

But

$$\phi_i(x_j) = \begin{cases} 0 & i \ne j, \\ 1 & i = j, \end{cases}$$

so $c_j \phi_j(x_j) = c_j = 0$. Hence the functions are linearly independent.

7. Suppose $\phi(x) = \sum_{i=0}^{n+1} c_i \phi_i(x) = 0$ for all x in $[0,1]$. At the nodes x_i, $i = 0, \ldots, n+1$, we have

$$\phi_0(x_i) = \begin{cases} 1/4, & \text{if } i = 1 \\ 0, & \text{otherwise;} \end{cases}$$

$$\phi_1(x_i) = \begin{cases} 1, & \text{if } i = 1 \\ 1/4, & \text{if } i = 2 \\ 0, & \text{otherwise;} \end{cases}$$

$$\phi_n(x_i) = \begin{cases} 1, & \text{if } i = n \\ 1/4, & \text{if } i = n-1 \\ 0, & \text{otherwise;} \end{cases}$$

$$\phi_{n+1}(x_i) = \begin{cases} 1/4, & \text{if } i = n \\ 0, & \text{otherwise;} \end{cases}$$

and for $j = 2, 3, \ldots, n-1$,

$$\phi_j(x_i) = \begin{cases} 1, & \text{if } i = j \\ 1/4, & \text{if } i = j-1 \text{ or } i = j+1 \\ 0, & \text{otherwise.} \end{cases}$$

Thus,

$$0 = \phi(x_1) = \frac{1}{4}c_0 + c_1 + \frac{1}{4}c_2$$

$$0 = \phi(x_2) = \frac{1}{4}c_1 + c_2 + \frac{3}{4}c_3$$

$$\vdots$$

$$0 = \phi(x_{n-1}) = \frac{1}{4}c_{n-2} + c_{n-1} + \frac{1}{4}c_n$$

$$0 = \phi(x_n) = \frac{1}{4}c_{n-1} + c_n + \frac{1}{4}c_{n+1}.$$

Since $\phi'(0) = \phi'(1) = 0$, we have

$$0 = \frac{3}{h}c_0 + \frac{1.5}{h}c_1, \quad \text{so} \quad 0 = 3c_0 + 1.5c_1$$

and

$$0 = -\frac{1.5}{h}c_n - \frac{3}{h}c_{n+1}, \quad \text{so} \quad 0 = 1.5c_n + 3c_{n+1}.$$

Thus,

$$\begin{bmatrix} 3 & 1.5 & 0 & \cdots & \cdots & \cdots & \cdots & 0 \\ 0.25 & 1 & 0.25 & \ddots & & & & \vdots \\ 0 & 0.25 & 1 & 0.25 & \ddots & & & \vdots \\ \vdots & \ddots & \ddots & \ddots & \ddots & \ddots & & \vdots \\ \vdots & & \ddots & \ddots & \ddots & \ddots & \ddots & \vdots \\ & & & \ddots & 0.25 & 1 & 0.25 & 0 \\ \vdots & & & & \ddots & 0.25 & 1 & 0.25 \\ 0 & \cdots & \cdots & \cdots & \cdots & 0 & 1.5 & 3 \end{bmatrix} \begin{bmatrix} c_0 \\ c_1 \\ \vdots \\ \vdots \\ \vdots \\ c_{n-1} \\ c_n \\ c_{n+1} \end{bmatrix} = \begin{bmatrix} 0 \\ 0 \\ \vdots \\ \vdots \\ \vdots \\ 0 \\ 0 \\ 0 \end{bmatrix},$$

130

which can be written as the linear system $Ac = 0$. The matrix A is strictly diagonally dominant and, hence, nonsingular. Thus, the only solution to the linear system is $c = 0$. Thus $\{\phi_0, \phi_1, \ldots, \phi_n, \phi_{n+1}\}$ is linearly independent.

8. Let $c = (c_1, \ldots, c_n)^t$ be any vector and let $\phi(x) = \sum_{j=1}^{n} c_j \phi_j(x)$. Then

$$
\begin{aligned}
c^t A c &= \sum_{i=1}^{n} \sum_{j=1}^{n} a_{ij} c_i c_j \\
&= \sum_{i=1}^{n} \sum_{j=i-1}^{i+1} a_{ij} c_i c_j \\
&= \sum_{i=1}^{n} \left[\int_0^1 \{ p(x) c_i \phi_i'(x) c_{i-1} \phi_{i-1}'(x) + q(x) c_i \phi_i(x) c_{i-1} \phi_{i-1}(x) \} \, dx \right. \\
&\quad + \int_0^1 \{ p(x) c_i^2 [\phi_i'(x)]^2 + q(x) c_i^2 [\phi_i'(x)]^2 \} \, dx \\
&\quad \left. + \int_0^1 \{ p(x) c_i \phi_i'(x) c_{i+1} \phi_{i+1}'(x) + q(x) c_i \phi_i(x) c_{i+1} \phi_{i+1}(x) \} \, dx \right] \\
&= \int_0^1 \{ p(x) [\phi'(x)]^2 + q(x) [\phi(x)]^2 \} \, dx.
\end{aligned}
$$

So $c^t A c \geq 0$ with equality only if $c = 0$.

10. $a_{i,i} = 2, \quad 1 \leq i \leq n; \quad a_{i,i-1} = -1, \quad 2 \leq i \leq n; \quad a_{i,i+1} = -1, \quad 1 \leq i \leq n-1.$

12. $c_1 = -0.03197519, \quad c_2 = -0.05488716, \quad c_3 = -0.06959856,$
$c_4 = -0.07688995, \quad c_5 = -0.07746753, \quad c_6 = -0.07197026,$
$c_7 = -0.06097626, \quad c_8 = -0.04500857, \quad c_9 = -0.02454043$

14. For $c = (c_0, c_1, \ldots, c_{n+1})^t$ and $\phi(x) = \sum_{i=0}^{n+1} c_i \phi_i(x)$,

$$
c^t A c = \int_0^1 p(x) [\phi'(x)]^2 + q(x) [\phi(x)]^2 \, dx.
$$

But $p(x) > 0$ and $q(x) [\phi(x)]^2 \geq 0$, so $c^t A c \geq 0$ and can be zero only if $\phi'(x) = 0$ on $[0, 1]$. However, $\{\phi_0', \phi_1', \ldots, \phi_{n+1}'\}$ is linearly independent, so $c^t A c = 0$ if and only if $c = 0$.

Chapter 12
Numerical Solutions to Partial-Differential Equations

Exercise Set 12.2 (PAGE 619)

2.

i	j	x_i	y_j	$w_{i,j}$
1	1	1/3	1/3	0.00617284
1	2	1/3	2/3	0.00617284
2	1	2/3	1/3	0.0493827
2	2	2/3	2/3	0.0493827

4. a)

x_i	y_j	$w_{i,j}(h = 0.2)$	$w_{i,j}(h = 0.1)$	$w_{i,j}(h = 0.05)$
0.2	0.4	0.07998843	0.07994399	0.07767125
0.4	0.6	0.2399813	0.2399094	0.2362279
0.8	0.8	0.6399942	0.6399687	0.6386266

x_i	y_j	Extrapolated
0.2	0.4	0.07671264
0.4	0.6	0.2346751
0.8	0.8	0.6380605

b)

x_i	y_j	$w_{i,j}(h = 0.2)$	$w_{i,j}(h = 0.1)$	$w_{i,j}(h = 0.05)$
0.2	0.4	0.8391384	0.8076865	0.7980150
0.4	0.6	1.857762	1.795605	1.777020
0.8	0.8	3.833466	3.781883	3.767512

x_i	y_j	Extrapolated
0.2	0.4	0.7946304
0.4	0.6	1.770554
0.8	0.8	3.762591

5. Let $N = (n-1)(m-1)$. Set up the $N \times N$ matrix A and the N dimensional column vector $\mathbf{b}$ using the correspondence

$$l \leftrightarrow i + (m - 1 - j)(n - 1)$$

as follows:

$A = (a_{l,l_1})$ where

$$a_{l,l_1} = \begin{cases} 2 + 2h^2/k^2, & \text{if } l = l_1 \\ -1, & \text{if } l_1 = l - 1 \text{ or } l_1 = l + 1 \text{ and } 1 \le l_1 \le N \\ -h^2/k^2, & \text{if } l_1 = l - (n-1) \text{ or } l_1 = l + (n-1) \text{ and } 1 \le l_1 \le N \end{cases}$$

and

$$b_l = -h^2 f(x_i, y_j) \quad \text{where } l = i + (m - 1 - j)(n - 1).$$

If

$$l - 1 = 0, \text{ add } g(x_{i-1}, y_j) \text{ to } b_l$$
$$l + 1 = 0, \text{ add } g(x_{i+1}, y_j) \text{ to } b_l$$
$$l - (n-1) = 0, \text{ add } h^2/k^2 g(x_i, y_{j+1}) \text{ to } b_l$$
$$l + (n-1) = 0, \text{ add } h^2/k^2 g(x_i, y_{j-1}) \text{ to } b_l.$$

Apply Choleski's Method to solve the linear system

$$A\mathbf{w} = \mathbf{b}$$

by factoring A into LL^t where L is lower triangular and solving in turn

$$L\mathbf{z} = \mathbf{b} \quad \text{and} \quad L^t\mathbf{w} = \mathbf{z}.$$

The method works well only if N is relatively small, substantially less than 100.

6. (3a)

i	j	x_i	y_j	$w_{i,j}$
3	3	0.3	0.3	0.09
3	7	0.3	0.7	0.21
7	3	0.7	0.3	0.21
7	7	0.7	0.7	0.49

(3b) Using $h = k = 0.1$:

i	j	x_i	y_j	$w_{i,j}$
3	3	0.3	0.3	1.103967
3	7	0.3	0.7	1.103967
7	3	0.7	0.3	3.839009
7	7	0.7	0.7	3.839009

(3c)

i	j	x_i	y_j	$w_{i,j}$
4	3	0.8	0.3	1.271453
4	7	0.8	0.7	1.750950
8	3	1.6	0.3	1.616795
8	7	1.6	0.7	3.065922

(3d)

i	j	x_i	y_j	$w_{i,j}$
2	1	1.256637	0.3141593	0.2951849
2	3	1.256637	0.9424778	0.1830810
4	1	2.513274	0.3141593	−0.7721951
4	3	2.513274	0.9424778	−0.4785174

7. To incorporate the SOR method, make the following changes to Algorithm 12.1:

STEP 1 Set

$$h = (b - a)/n;$$
$$k = (d - c)/m;$$
$$\omega = 4 / \left(2 + \sqrt{4 - [\cos \pi/m]^2 - [\cos \pi/n]^2} \right);$$
$$\omega_0 = 1 - \omega.$$

134

In each of STEPS 7, 8, 9, 11, 12, 13, 14, 15, and 16 after

        Set $z = \ldots$

insert

        Set $E = w_{\alpha,\beta} - z$;

           if $(\ |E| > NORM)$ then set $NORM = |E|$;

        Set $w_{\alpha,\beta} = \omega_0 E + z$.

where α and β depend on which STEP is being changed.

8. The answers are the same as in Exercise 3. A tolerance of 10^{-4} required the following number of iterations:

Algorithm 12.1	(Gauss-Seidel)	SOR
(3a)	61	20
(3b)	284	45
(3c)	80	21
(3d)	18	10

10.

i	j	x_i	y_j	$w_{i,j}$
5	9	2.0	3.0	5.957716
8	3	3.2	1.0	7.915441
10	9	4.0	3.0	4.678240
12	12	4.8	4.0	2.059610

Exercise Set 12.3 (PAGE 632)

2. a)

i	j	x_i	t_j	$w_{i,j}$
1	1	0.5	0.05	0.628848
2	1	1.0	0.05	0.889326
3	1	1.5	0.05	0.628848
1	2	0.5	0.1	0.559251
2	2	1.0	0.1	0.790901
3	2	1.5	0.1	0.559252

135

b)

i	j	x_i	t_j	$w_{i,j}$
1	1	1/3	0.05	1.591825
2	1	2/3	0.05	−1.591825
1	2	1/3	0.1	1.462951
2	2	2/3	0.1	−1.462951

4. a) For $h = 0.1$ and $k = 0.01$:

i	j	x_i	t_j	$w_{i,j}$
4	50	0.4	0.5	1.863488×10^{-7}
10	50	1.0	0.5	2.414952×10^{-7}
17	50	1.7	0.5	2.969643×10^{-8}

For $h = 0.1$ and $k = 0.005$:

i	j	x_i	t_j	$w_{i,j}$
4	100	0.4	0.5	1.526075×10^{-7}
10	100	1.0	0.5	2.449084×10^{-7}
17	100	1.7	0.5	9.256785×10^{-8}

b) For $h = 0.1$ and $k = 0.01$:

i	j	x_i	t_j	$w_{i,j}$
4	50	0.4	0.5	0.1797665
10	50	1.0	0.5	0.3058325
17	50	1.7	0.5	0.1388477

c) For $h = 0.2$ and $k = 0.04$:

i	j	x_i	t_j	$w_{i,j}$
5	10	1	0.4	1.317964
10	10	2	0.4	0.9054714
15	10	3	0.4	−0.03743314

d) For $h = 0.1$ and $k = 0.04$:

i	j	x_i	t_j	$w_{i,j}$
3	10	0.3	0.4	0.5482691
5	10	0.5	0.4	0.6776979
7	10	0.7	0.4	0.5482691

6. (3a) Using $h = 0.1$ and $k = 0.01$ leads to meaningless results. Using $h = 0.1$ and $k = 0.005$ again gives meaningless answers. Letting $h = 0.4$ and $k = 0.005$ yields the following:

i	j	x_i	t_j	$w_{i,j}$
1	100	0.4	0.5	-165.405
2	100	0.8	0.5	267.613
3	100	1.2	0.5	-267.613
4	100	1.6	0.5	165.405

(3b) Using $h = 0.1$ and $k = 0.0005$ leads to meaningless results. Letting $h = 0.4$ and $k = 0.005$ yields the following:

i	j	x_i	t_j	$w_{i,j}$
1	100	0.4	0.5	-0.0346524
2	100	0.8	0.5	0.0427262
3	100	1.2	0.5	0.0427262
4	100	1.6	0.5	-0.0346524

(3c)

i	j	x_i	t_j	$w_{i,j}$
4	10	0.8	0.4	1.14063
8	10	1.6	0.4	1.23160
12	10	2.4	0.4	0.472676
16	10	3.2	0.4	-0.0873302

(3d)

i	j	x_i	t_j	$w_{i,j}$
2	10	0.2	0.4	0.379460
4	10	0.4	0.4	0.613979
6	10	0.6	0.4	0.613979
8	10	0.8	0.4	0.379460

7. We have

$$a_{11}v_1^{(i)} + a_{12}v_2^{(i)} = (1 - 2\lambda)\sin\frac{i\pi}{m} + \lambda\sin\frac{2\pi i}{m}$$

and

$$\mu_i v_1^{(i)} = \left[1 - 4\lambda(\sin\frac{i\pi}{2m})^2\right]\sin\frac{i\pi}{m}$$

$$= \left[1 - 4\lambda(\sin\frac{i\pi}{2m})^2\right]\left(2\sin\frac{i\pi}{2m}\cos\frac{i\pi}{2m}\right)$$

$$= 2\sin\frac{i\pi}{2m}\cos\frac{i\pi}{2m} - 8\lambda(\sin\frac{i\pi}{2m})^3\cos\frac{i\pi}{2m}.$$

However,

$$(1 - 2\lambda)\sin\frac{i\pi}{m} + \lambda\sin\frac{2\pi i}{m} = 2(1 - 2\lambda)\sin\frac{i\pi}{2m}\cos\frac{i\pi}{2m} + 2\lambda\sin\frac{i\pi}{m}\cos\frac{i\pi}{m}$$

$$= 2(1 - 2\lambda)\sin\frac{i\pi}{2m}\cos\frac{i\pi}{2m}$$

$$+ 2\lambda\left[2\sin\frac{i\pi}{2m}\cos\frac{i\pi}{2m}\right]\left[1 - 2(\sin\frac{i\pi}{2m})^2\right]$$

$$= 2\sin\frac{i\pi}{2m}\cos\frac{i\pi}{2m} - 8\lambda\cos\frac{i\pi}{2m}\left[\sin\frac{i\pi}{2m}\right]^3.$$

Thus,

$$a_{11}v_1^{(i)} + a_{12}v_2^{(i)} = \mu_i v_1^{(i)}.$$

Further,

$$a_{j,j-1}v_{j-1}^{(i)} + a_{j,j}v_j^{(i)} + a_{j,j+1}v_{j+1}^{(i)} = \lambda\sin\frac{i(j-1)\pi}{m} + (1 - 2\lambda)\sin\frac{ij\pi}{m} + \lambda\sin\frac{i(j+1)\pi}{m}$$

$$= \lambda\left(\sin\frac{ij\pi}{m}\cos\frac{i\pi}{m} - \sin\frac{i\pi}{m}\cos\frac{ij\pi}{m}\right) + (1 - 2\lambda)\sin\frac{ij\pi}{m}$$

$$+ \lambda\left(\sin\frac{ij\pi}{m}\cos\frac{i\pi}{m} + \sin\frac{i\pi}{m}\cos\frac{ij\pi}{m}\right)$$

$$= \sin\frac{ij\pi}{m} - 2\lambda\sin\frac{ij\pi}{m} + 2\lambda\sin\frac{ij\pi}{m}\cos\frac{i\pi}{m}$$

$$= \sin\frac{ij\pi}{m} + 2\lambda\sin\frac{ij\pi}{m}\left(\cos\frac{i\pi}{m} - 1\right)$$

and

$$\mu_i v_j^{(i)} = \left[1 - 4\lambda(\sin\frac{i\pi}{2m})^2\right]\sin\frac{ij\pi}{m}$$

$$= \left[1 - 4\lambda(\frac{1}{2} - \frac{1}{2}\cos\frac{i\pi}{m})\right]\sin\frac{ij\pi}{m}$$

$$= \left[1 + 2\lambda(\cos\frac{i\pi}{m} - 1)\right]\sin\frac{ij\pi}{m},$$

so

$$a_{j,j-1}^{(i)}v_{j-1}^{(i)} + a_{j,j}v_j^{(i)} + a_{j,j+1}v_j^{(i)} = \mu_i v_j^{(i)}.$$

Similarly, $a_{m-2,m-1}v_{m-2}^{(i)} + a_{m-1,m-1}v_{m-1}^{(i)} = \mu_i v_{m-1}^{(i)}$, so $A\mathbf{v}^{(i)} = \mu_i \mathbf{v}^{(i)}$.

8. We have

$$a_{11}^{(i)}v_1^{(i)} + a_{12}^{(i)}v_2^{(i)} = (1 + 2\lambda)\sin\frac{i\pi}{m} - \lambda\sin\frac{i\pi}{2m}$$

$$= (1 + 2\lambda)\sin\frac{i\pi}{m} - 2\lambda\sin\frac{i\pi}{m}\cos\frac{i\pi}{m}$$

$$= \sin\frac{i\pi}{m}\left[1 + 2\lambda(1 - \cos\frac{i\pi}{m})\right]$$

and

$$\mu_i v_1^{(i)} = \left(1 + 4\lambda[\sin\frac{i\pi}{2m}]^2\right)\sin\frac{i\pi}{m}$$

$$= \left[1 + 2\lambda(1 - \cos\frac{i\pi}{m})\right]\sin\frac{i\pi}{m} = a_{11}^{(i)}v_1^{(i)} + a_{12}^{(i)}v_2^{(i)}.$$

In general,

$$a_{j,j-1}v_{j-1}^{(i)} + a_{j,j}v_j^{(i)} + a_{j,j+1}v_{j+1}^{(i)} = -\lambda\sin\frac{i(j-1)\pi}{m}$$

$$+ (1 + 2\lambda)\sin\frac{ij\pi}{m} - \lambda\sin\frac{i(j+1)\pi}{m}$$

$$= -\lambda\left(\sin\frac{ij\pi}{m}\cos\frac{i\pi}{m} - \sin\frac{i\pi}{m}\cos\frac{ij\pi}{m}\right)$$

$$+ (1 + 2\lambda)\sin\frac{ij\pi}{m}$$

$$- \lambda\left(\sin\frac{ij\pi}{m}\cos\frac{i\pi}{m} - \sin\frac{i\pi}{m}\cos\frac{ij\pi}{m}\right)$$

$$= -2\lambda\sin\frac{ij\pi}{m}\cos\frac{i\pi}{m} + (1 + 2\lambda)\sin\frac{ij\pi}{m}$$

$$= \left[1 + 2\lambda(1 - \cos\frac{i\pi}{m})\right]\sin\frac{ij\pi}{m}$$

$$= \mu_i v_j^{(i)}.$$

Similarly,

$$a_{m-2,m-1}v_{m-2}^{(i)} + a_{m-1,m-1}v_{m-1}^{(i)} = \mu_i v_{m-1}^{(i)}.$$

Thus, $A\mathbf{v}^{(i)} = \mu_i \mathbf{v}^{(i)}$.

9. To modify Algorithm 12.2, change the following:
   STEP 7 Set
   $$t = jk;$$
   $$z_1 = (w_1 + kF(h))/l_1.$$
   STEP 8 For $i = 2, \ldots, m-1$ set
   $$z_i = (w_i + kF(ih) + \lambda z_{i-1})/l_i.$$

   To modify Algorithm 12.3, change the following:
   STEP 7 Set
   $$t = jk;$$
   $$z_1 = \left[(1-\lambda)w_1 + \tfrac{\lambda}{2}w_2 + kF(h)\right]/l_1.$$
   STEP 8 For $i = 2, \ldots, m-1$ set
   $$z_i = \left[(1-\lambda)w_i + \tfrac{\lambda}{2}\left(w_{i+1} + w_{i-1} + z_{i-1}\right) + kF(ih)\right]/l_i.$$

10. For modified Algorithm 12.2:

i	j	x_i	t_j	$w_{i,j}$
3	25	0.3	0.25	0.2883455
5	25	0.5	0.25	0.3468410
8	25	0.8	0.25	0.2169213

For modified Algorithm 12.3:

i	j	x_i	t_j	$w_{i,j}$
3	25	0.3	0.25	0.2798727
5	25	0.5	0.25	0.3363676
8	25	0.8	0.25	0.2107654

11. To modify Algorithm 12.2, change the following:

STEP 7 Set

$$t = jk;$$
$$w_0 = \phi(t);$$
$$z_1 = (w_1 + \lambda w_0)/l_1.$$
$$w_m = \psi(t).$$

STEP 8 For $i = 2, \ldots, m-2$ set

$$z_i(w_i + \lambda z_{i-1})/l_i;$$
$$z_{m-1} = (w_{m-1} + \lambda w_m + \lambda z_{m-2})/l_{m-1}.$$

STEP 11 OUTPUT (t);

For $i = 0, \ldots, m$ set

$$x = ih;$$
$$\text{OUTPUT } (x, w_i).$$

To modify Algorithm 12.3, change the following:

STEP 1 Set

$$h = l/m;$$
$$k = T/N;$$
$$\lambda = \alpha^2 k/h^2;$$
$$w_m = \psi(0);$$
$$w_0 = \phi(0);$$

STEP 7 Set

$$t = jk;$$
$$z_1 = [(1-\lambda)w_1 + \tfrac{\lambda}{2}w_2 + \tfrac{\lambda}{2}w_0 + \tfrac{\lambda}{2}\phi(t)]/l_1;$$
$$w_0 = \phi(t).$$

STEP 8 For $i = 2, \ldots, m-2$ set

$$z_i = [(1-\lambda)w_i + \tfrac{\lambda}{2}(w_{i+1} + w_{i-1} + z_{i-1})]/l_i;$$

Set

$$z_{m-1} = [(1-\lambda)w_{m-1} + \tfrac{\lambda}{2}(w_m + w_{m-2} + z_{m-2} + \psi(t))]/l_{m-1};$$
$$w_m = \psi(t).$$

STEP 11 OUTPUT (t);

For $i = 0, \ldots, m$ set $x = ih$;

OUTPUT (x, w_i).

12.

i	j	x_i	t_j	$w_{i,j}$(Algorithm 12.3)	$w_{i,j}$(Algorithm 12.2)
3	10	0.3	0.225	1.223272	1.207730
6	10	0.75	0.225	1.862347	1.836564
10	10	1.35	0.225	0.7010836	0.6928342

14.

i	j	x_i	t_j	$w_{i,j}$
2	10	200	5	1.478828×10^7
5	10	500	5	4.334451×10^6
8	10	800	5	1.478828×10^7

Exercise Set 12.4 (Page 642)

2.

i	j	x_i	t_j	$w_{i,j}$
1	2	0.0	0.5	0
2	2	0.25	0.5	-1.740383×10^{-8}
3	2	0.5	0.5	0

4. For $h = 0.1$ and $k = 0.05$:

i	j	x_i	t_j	$w_{i,j}$
2	10	0.2	0.5	0.0489677
5	10	0.5	0.5	0.0833088
7	10	0.7	0.5	0.0673983

For $h = 0.05$ and $k = 0.1$:

i	j	x_i	t_j	$w_{i,j}$
4	5	0.2	0.5	0.0905112
10	5	0.5	0.5	0.1539869
14	5	0.7	0.5	0.1245780

For $h = 0.05$ and $k = 0.05$:

i	j	x_i	t_j	$w_{i,j}$
4	10	0.2	0.5	0.0462597
10	10	0.5	0.5	0.0787017
14	10	0.7	0.5	0.0636710

6.

i	j	x_i	t_j	$w_{i,j}$
2	5	0.2	0.5	-1
5	5	0.5	0.5	0
7	5	0.7	0.5	1

8. a) For voltage V:

i	j	x_i	t_j	$w_{i,j}$
5	2	50	0.2	77.782
12	2	120	0.2	104.62
18	2	180	0.2	33.992
5	5	50	0.5	77.782
12	5	120	0.5	104.62
18	5	180	0.5	33.992

For current i:

i	j	x_i	t_j	$w_{i,j}$
5	2	50	0.2	3.88909
12	2	120	0.2	-1.69959
18	2	180	0.2	-5.23081
5	5	50	0.5	3.88908
12	5	120	0.5	-1.6959
18	5	180	0.5	-5.23081

Exercise Set 12.5 (Page 656)

2. With $E_1 = (0.25, 0.75), E_2 = (0, 1), E_3 = (0.5, 0.5), E_4 = (0, 0.5), E_5 = (0, 0.75)$, and $E_6 = (0.25, 0.5)$ the following results are obtained:

i	j	$a_j^{(i)}$	$b_j^{(i)}$	$c_j^{(i)}$		i	j	$a_j^{(i)}$	$b_j^{(i)}$	$c_j^{(i)}$
1	1	0	4	0		3	1	0	4	0
1	2	−3	0	4		3	2	3	0	−4
1	3	4	−4	−4		3	3	−2	4	4
2	1	−2	0	4		4	1	−2	0	4
2	2	−1	4	0		4	2	1	−4	0
2	3	4	−4	−4		4	3	2	4	−4

So $\gamma_1 = 0.3461969, \gamma_2 = 0, \gamma_3 = 1.0, \gamma_4 = 0, \gamma_5 = 0, \gamma_6 = 0.5.$

4. $K = 8, N = 22, M = 32, n = 25, m = 25, NL = 16$;

$\gamma_1 = -0.489695$ $\gamma_9 = -0.489695$ $\gamma_{17} = 0.058157$

$\gamma_2 = 0.016323$ $\gamma_{10} = -1.06913$ $\gamma_{18} = 0.058157$

$\gamma_3 = 0.524240$ $\gamma_{11} = -0.684308$ $\gamma_{19} = 0.752868$

$\gamma_4 = 0.016325$ $\gamma_{12} = 0.058157$ $\gamma_{20} = -0.684310$

$\gamma_5 = 0.008685$ $\gamma_{13} = 0.752868$ $\gamma_{21} = 0.962801$

$\gamma_6 = 0.016324$ $\gamma_{14} = 0.962799$ $\gamma_{22} = 0.752870$

$\gamma_7 = 0.524242$ $\gamma_{15} = -0.684308$ $\gamma_{23} = 0.058157$

$\gamma_8 = 0.016325$ $\gamma_{16} = 0.752869$ $\gamma_{24} = -0.684312$

$\gamma_{25} = -1.06913$

$u(0.125, 0.125) \simeq 0.270284, u(0.125, 0.25) \simeq -0.238595$

$u(0.25, 0.125) \simeq -0.238595, u(0.25, 0.25) \simeq 0.016323$

```
                    BISECTION ALGORITHM   2.1


C     TO FIND A SOLUTION TO F(X)=0 GIVEN THE CONTINOUS FUNCTION
C     F ON THE INTERVAL <A,B>, WHERE F(A) AND F(B) HAVE
C     OPPOSITE SIGNS:
C
C     INPUT:    ENDPOINTS A,B; TOLERANCE TOL;
C               MAXIMUM INTERATIONS NO.
C
C     OUTPUT:   APPROXIMATE SOLUTION P OR A
C               MESSAGE THAT THE ALGORITHM FAILS.
C
C     DEFINE F
      F(X)=X**3+4*X**2-10
C     INPUT:  DATA FOLLOWS THE END STATEMENT
      READ(5,1) A,B,TOL,NO
C     STEP 1
      I=1
C     STEP 2
      WHILE(I.LE.NO) DO
C          STEP 3
C          COMPUTE P(I)
           P=A+(B-A)/2
C          STEP 4
           IF( F(P).LE.1.0E-20 .OR. (B-A)/2 .LT. TOL) THEN DO
C          PROCEDURE COMPLETED SUCCESSFULLY
                WRITE(6,2) P, I, TOL
                WRITE(6,4)
                STOP
           END IF
C          STEP 5
           I=I+1
C          STEP 6
C          COMPUTE A(I) AND B(I)
           IF( F(A)*F(P) .GT. 0) THEN DO
                A=P
           ELSE DO
                B=P
           END IF
      END WHILE
C     STEP 7
C     PROCEDURE COMPLETED UNSUCCESSFULLY
      WRITE(6,3) NO
      WRITE(6,4)
```

145

```
      STOP
1     FORMAT(3(E15.8,1X),I2)
2     FORMAT('1','THE APPROXIMATE SOLUTION IS',1X
      *,E15.8,1X,'AFTER',1X,I2,1X,'ITERATIONS, WITH TOLERANCE'
      * ,1X,E15.8)
3     FORMAT('1','THE METHOD FAILS AFTER',1X,I2,1X,'
      *ITERATIONS')
4     FORMAT('1')
      END
$ENTRY


      1.0E+00         2.0E+00          5.0E-04 20
/*
```

FIXED-POINT ALGORITHM   2.2

```
C     TO FIND A SOLUTION TO P=G(P) GIVEN AN
C     INITIAL APPROXIMATION P0:
C
C     INPUT:    INITIAL APPROXIMATION P0; TOLERANCE TOL;
C               MAXIMUM NUMBER OF ITERATIONS N0.
C
C     OUTPUT:   APPROXIMATE SOLUTION P OR MESSAGE THAT
C               THE METHOD FAILS.
C
C     DEFINE FUNCTION G
      G(X)=SQRT(10/(4+X))
C     INPUT:  DATA FOLLOWS THE END STATEMENT
      READ(5,1) P0,TOL,N0
C     STEP 1
      I=1
C     STEP 2
      WHILE( I .LE. N0 ) DO
C          STEP 3
C          COMPUTE P(I)
           P=G(P0)
C          STEP 4
           IF( ABS(P-P0) .LT. TOL ) THEN DO
```

146

```
C           PROCEDURE COMPLETED SUCCESSFULLY
                  WRITE(6,2) P, I, TOL
                  WRITE(6,4)
                  STOP
              END IF
C         STEP 5
              I=I+1
C         STEP 6
C         UPDATE P0
              P0=P
      END WHILE
C     STEP 7
C     PROCEDURE COMPLETED UNSUCCESSFULLY
      WRITE(6,3) NO
      WRITE(6,4)
      STOP
1     FORMAT(2(E15.8,1X),I2)
2     FORMAT('1','APPROXIMATE SOLUTION IS',1X,E15.8,1X
     *,'AFTER',1X,I2,1X,'ITERATIONS, WITH TOLERANCE',1X,E15.8)
3     FORMAT('1','METHOD FAILS AFTER',1X,I2,1X,'ITERATIONS')
4     FORMAT('1')
      END
$ENTRY
          1.5E+00          5.0E-04 25
/*
```

NEWTON-RAPHSON ALGORITHM   2.3

```
C     TO FIND A SOLUTION TO F(X)=0 GIVEN AN
C     INITIAL APPROXIMATION P0 ASSUMING F'(X) EXISTS:
C
C     INPUT:    INITIAL APPROXIMATION P0;  TOLERANCE TOL;
C               MAXIMUM NUMBER OF ITERATIONS NO.
C
C     OUTPUT:   APPROXIMATE SOLUTION P OR A MESSAGE
C               THAT THE ALGORITHM FAILS.
C
C     DEFINE FUNCTIONS F AND F' (DENOTED FP)
      F(X)=COS(X)-X
      FP(X)=-SIN(X)-1
```

```fortran
C      INPUT:  DATA FOLLOWS THE END STATEMENT
       READ(5,1) P0,TOL,N0
C      STEP 1
       I=1
C      STEP 2
       WHILE ( I .LE. N0 ) DO
C          STEP 3
C          COMPUTE P(I)
           P=P0-F(P0)/FP(P0)
C          STEP 4
           IF( ABS(P-P0) .LT. TOL ) THEN DO
C          PROCEDURE COMPLETED SUCCESSFULLY
               WRITE(6,2) P,I,TOL
               WRITE(6,4)
               STOP
           END IF
C          STEP 5
           I=I+1
C          STEP 6
C          UPDATE P0
           P0=P
       END WHILE
C      STEP 7
C      PROCEDURE COMPLETED UNSUCCESSFULLY
       WRITE(6,3) N0
       WRITE(6,4)
       STOP
1      FORMAT(2(E15.8,1X),I2)
2      FORMAT('1','APPROXIMATE SOLUTION IS',1X,E15.8,1X,'AFTER',
      *1X,I2,1X,'ITERATIONS, WITH TOLERANCE',1X,E15.8)
3      FORMAT('1','METHOD FAILS AFTER',1X,I2,1X,'ITERATONS')
4      FORMAT('1')
       END
$ENTRY
  .78539816E+00          0.5E-03 15
/*
```

```
                    SECANT ALGORITHM   2.4


C     TO FIND A SOLUTION TO THE EQUATION F(X)=0
C     GIVEN INITIAL APPROXIMATIONS P0 AND P1:
C
C     INPUT:    INITIAL APPROXIMATION P0, P1; TOLERANCE TOL;
C               MAXIMUM NUMBER OF ITERATIONS N0.
C
C     OUTPUT:   APPROXIMATE SOLUTION P OR MESSAGE THAT THE
C               ALGORITHM FAILS.
C
C     DEFINE FUNCTION F
      F(X)=COS(X)-X
C     INPUT:  DATA FOLLOWS THE END STATEMENT
      READ(5, 1) P0, P1, TOL, N0
C     STEP 1
      I=2
      Q0=F(P0)
      Q1=F(P1)
C     STEP 2
      WHILE( I .LE. N0 ) DO
C         STEP 3
C         COMPUTE P(I)
          P=P1-Q1*(P1-P0)/(Q1-Q0)
C         STEP 4
          IF( ABS(P-P1) .LT. TOL ) THEN DO
              WRITE(6, 2) P, I, TOL
              WRITE(6, 4)
              STOP
          END IF
C         STEP 5
          I=I+1
C         STEP 6
C         UPDATE P0, Q0, P1, Q1
          P0=P1
          Q0=Q1
          P1=P
          Q1=F(P)
      END WHILE
C     STEP 7
C     PROCEDURE COMPLETED UNSUCCESSFULLY
      WRITE(6, 3) N0
      WRITE(6, 4)
```

```
      STOP
1     FORMAT(3(E15.8, 1X), I2)
2     FORMAT('1', 'THE APPROXIMATE SOLUTION IS', 1X, E15.8, 1X,
     *'AFTER', 1X, I2, 1X, 'ITERATIONS, WITH TOLERANCE', 1X, E15.8)
3     FORMAT('1', 'THE METHOD FAILS AFTER', 1X, I2, 1X, 'ITERATIONS')
4     FORMAT('1')
      END
$ENTRY
      0.5E+00    .78539816E+00          0.5E-04 15
/*
```

STEFFENSEN'S ALGORITHM   2.5

```
C     TO FIND A SOLUTION TO G(X) = X
C     GIVEN AN INITIAL APPROXIMATION P0:
C
C     INPUT:     INITIAL APPROXIMATION P0, TOLERANCE TOL,
C                MAXIMUM NUMBER OF ITERATIONS N0.
C
C     OUTPUT:    APPROXIMATE SOLUTION P OR MESSAGE THAT
C                THE METHOD FAILS.
C
C     DEFINE FUNCTION G
      G(X)=SQRT(10/(X+4))
C     INPUT:  DATA FOLLOWS THE END STATEMENT
      READ(5, 1) P0, TOL, N0
C     STEP 1
      I = 1
C     STEP 2
      WHILE( I .LE . N0 ) DO
C          STEP 3
C          COMPUTE P(1) WITH SUPERSCRIPT (I-1)
           P1 = G(P0)
C          COMPUTE P(2) WITH SUPERSCRIPT (I-1)
           P2 = G(P1)
           DIF = (P1-P0)*(P1-P0)/(P2-2*P1+P0)
C          COMPUTE P(0) WITH SUPERSCRIPT (I)
           P = P0-DIF
C          STEP 4
```

```
              IF( ABS(DIF) .LT. TOL ) THEN DO
C          PROCEDURE COMPLETED SUCCESSFULLY
                  WRITE(6,2) P,I,TOL
                  WRITE(6,4)
                  STOP
              END IF
C          STEP 5
              I = I+1
C          STEP 6
C          UPDATE P0
              P0=P
       END WHILE
C    STEP 7
C    PROCEDURE COMPLETED UNSUCCESSFULLY
       WRITE(6,3) N0
       WRITE(6,4)
       STOP
1      FORMAT( 2(E15.8,1X), I2)
2      FORMAT('1','THE APPROXIMATE SOLUTION IS',1X,E15.8,1X,
      *'AFTER',1X,I2,1X,'ITERATIONS, WITH TOLERANCE',1X,E15.8)
3      FORMAT('1','THE METHOD FAILS AFTER',1X,I2,1X,'ITERATIONS')
4      FORMAT('1')
       END
$ENTRY
         1.5E+00          .5E-04 15
   /*
```

HORNER'S ALGORITHM   2.6

```
C    TO EVALUATE THE POLYNOMIAL
C    P(X) = A(N)*X**N + A(N-1)*X**(N-1) + ... + A(1)*X + A(0)
C    AND ITS DERIVATIVE P'(X) AT X = X0;
C
C    INPUT:   DEGREE N; COEFFICIENTS AA(1),AA(2),...,AA(N+1);
C             VALUE OF X0.
C
C    OUTPUT:  Y = P(X0), Z = P'(X0).
C
```

```
C       DEFINE ARRAY AA
        DIMENSION AA(51)
C       INPUT:   DATA FOLLOWS THE END STATEMENT
        READ(5,1) N
        M=N+1
        READ(5,2) (AA(I),I=1,M)
        READ(5,3) X0
C       STEP 1
C       COMPUTE B(N) FOR P(X)
        Y=AA(N+1)
C       COMPUTE B(N-1) FOR Q(X)=P'(X)
        Z=AA(N+1)
        MM=N-1
C       STEP 2
        DO 10 I=1,MM
            J=N+1-I
C           COMPUTE B(J) FOR P(X)
            Y=Y*X0+AA(J)
C           COMPUTE B(J-1) FOR Q(X)
10      Z=Z*X0+Y
C       STEP 3
C       COMPUTE B(0) FOR P(X)
        Y=Y*X0+AA(1)
C       STEP 4
        WRITE(6,4)(AA(I),I=1,M)
        WRITE(6,5) X0,Y,X0,Z
        WRITE(6,6)
        STOP
1       FORMAT(I2)
2       FORMAT(5(E15.8,1X))
3       FORMAT(E15.8)
4       FORMAT('1',' COEFFICIENTS OF P ARE',6(E15.8,1X))
5       FORMAT('0',' P(',E15.8,') = ',E15.8,',  P''(',E15.8,') = ',E15.8)
6       FORMAT('1')
        END
$ENTRY
 4
        -4.0E+00        3.0E+00       -3.0E+00        0.0E+00
          2.0E+00       -2.0E+00
```

```
C      TO FIND A SOLUTION TO F(X) = 0 GIVEN THREE APPROXIMATIONS X0, X1
C      AND X2:
C
C      INPUT X0, X1, X2; TOLERANCE TOL; MAXIMUM NUMBER OF ITERATIONS N0.
C
C      OUTPUT APPROXIMATE SOLUTION P OR MESSAGE OF FAILURE.
C
C      THIS IMPLEMENTATION ALLOWS FOR A SWITCH TO COMPLEX ARITHMETIC
C      THE COEFFICIENTS ARE STORED IN THE VECTOR A, SO THE DIMENSION
C      OF A MAY HAVE TO BE CHANGED
       COMPLEX Z(4), G(4), CH(3), CDEL1(2), CDEL, CB, CD, CE
       DIMENSION H(3), F(4), X(4), DEL1(2), A(10)
C      INPUT:    DATA FOLLOWS THE END STATEMENT
       READ(5, 1)N, M, TOL
       N=N+1
       READ(5, 2)(A(I), I=1, N)
       WRITE(6, 3) (A(I), I=1, N)
       READ(5, 4) X(1), X(2), X(3)
C      EVALUATE F USING HORNER'S METHOD AND STORE IN VECTOR F
       DO 10 I=1, 3
          F(I)=A(N)
          DO 10 J=2, N
             K=N-J+1
10           F(I)=F(I)*X(I)+A(K)
       WRITE(6, 5) (X(I), F(I), I=1, 3)
C      VARIABLE ISW USED TO NOTE A SWITCH TO COMPLEX ARITHMETIC
C      ISW=0 MEANS REAL ARITHMETIC, AND ISW=1 MEANS COMPLEX ARITHMETIC
       ISW=0
C      STEP 1
       H(1)=X(2)-X(1)
       H(2)=X(3)-X(2)
       DEL1(1)=(F(2)-F(1))/H(1)
       DEL1(2)=(F(3)-F(2))/H(2)
       DEL=(DEL1(2)-DEL1(1))/(H(2)+H(1))
       I=2
C      STEP 2
       WHILE (I.LE.M) DO
C         STEPS 3-7 FOR REAL ARITHMETIC
          IF(ISW.EQ.0) THEN DO
C            STEP 3
             B=DEL1(2)+H(2)*DEL
```

```
                  D=B*B-4*F(3)*DEL
C                 TEST TO SEE IF NEED COMPLEX ARITHMETIC
                  IF(D.GE.0.0) THEN DO
C                    REAL ARITHMETIC/ TEST TO SEE IF STRAIGHT LINE
                     IF(ABS(DEL).LE.1.0E-20) THEN DO
C                       STRAIGHT LINE/ TEST TO SEE IF HORIZONTAL
                        IF(ABS(DEL1(2)).LE.1.0E-20) THEN DO
C                          HORIZONTAL LINE
                           WRITE(6,6)
                           STOP
                        ENDIF
C                       STRAIGHT LINE BUT NOT HORIZONTAL
                        X(4)=(F(3)-DEL1(2)*X(3))/DEL1(2)
                        H(3)=X(4)-X(3)
                     ELSE DO

C                       NOT STRAIGHT LINE
                        D=SQRT(D)
C                       STEP 4
                        E=B+D
                        IF(ABS(B-D).GT.ABS(E)) E=B-D
C                       STEP 5
                        H(3)=-2*F(3)/E
                        X(4)=X(3)+H(3)
                     ENDIF
C                    EVALUATE F(X(2))=F(4) BY HORNER'S METHOD
                     F(4)=A(N)
                     DO 30 J=2,N
                        K=N-J+1
30                   F(4)=F(4)*X(4)+A(K)
                     WRITE(6,8) I,X(4),F(4)
C                    STEP 6
                     IF(ABS(H(3)).LE.TOL) THEN DO
                        WRITE(6,9)
                        STOP
                     ENDIF
C                    STEP 7
                     DO 50 J=1,2
                        H(J)=H(J+1)
                        X(J)=X(J+1)
50                      F(J)=F(J+1)
                     X(3)=X(4)
                     F(3)=F(4)
                     DEL1(1)=DEL1(2)
                     DEL1(2)=(F(3)-F(2))/H(2)
                     DEL=(DEL1(2)-DEL1(1))/(H(2)+H(1))
                  ELSE DO
C                 SWITCH TO COMPLEX ARITHMETIC
                     ISW=1
```

154

```
             DO 60 J=1,3
                 Z(J)=X(J)
60               G(J)=F(J)
             DO 70 J=1,2
                 CDEL1(J)=DEL1(J)
70               CH(J)=H(J)
             CDEL=DEL
           ENDIF
         ENDIF
C        STEPS 3-7 COMPLEX ARITHMETIC
         IF(ISW.EQ.1) THEN DO
C          TEST IF STRAIGHT LINE
           IF(CABS(CDEL).LE.1.0E-20) THEN DO
C            STRAIGHT LINE/ TEST IF HORIZONTAL
             IF(CABS(CDEL1(1)).LE.1.0E-20) THEN DO
C              HORIZONTAL LINE
               WRITE(6,6)
               STOP
             ENDIF
C            STRAIGHT LINE BUT NOT HORIZONTAL
             Z(4)=(G(4)-CDEL1(2)*Z(3))/CDEL1(2)
             CH(3)=Z(4)-Z(3)
           ELSE DO
C            NOT STRAIGHT LINE
C              STEP 3
               CB=CDEL1(2)+CH(2)*CDEL
               CD=CB*CB-4*G(3)*CDEL
               CD=CSQRT(CD)
C              STEP 4
               CE=CB+CD
               IF(CABS(CB-CD).GT.CABS(CE)) CE=CB-CD
C              STEP 5
               CH(3)=-2*G(3)/CE
               Z(4)=Z(3)+CH(3)
           ENDIF
C          EVALUATE G(X(I))=G(4) BY HORNER'S METHOD
           G(4)=A(N)
           DO 80 J=2,N
               K=N-J+1
80         G(4)=G(4)*Z(4)+A(K)
           WRITE(6,11) I,Z(4),G(4)
C          STEP 6
           IF(CABS(CH(3)).LE.TOL) THEN DO
               WRITE(6,9)
               STOP
           ENDIF
C          STEP 7  CONTINUED
           DO 90 J=1,2
               CH(J)=CH(J+1)
```

155

```
                    Z(J)=Z(J+1)
90                    G(J)=G(J+1)
                Z(3)=Z(4)
                G(3)=G(4)
                CDEL1(1)=CDEL1(2)
                CDEL1(2)=(G(3)-G(2))/CH(2)
                CDEL=(CDEL1(2)-CDEL1(1))/(CH(2)+CH(1))
            ENDIF
C        STEP 7
            I=I+1
        ENDWHILE
C        STEP 8
        WRITE(6,12)
1        FORMAT(2I2,E15.8)
2        FORMAT(5E15.8)
3        FORMAT(8(1X,E15.8))
4        FORMAT(3E15.8)
5        FORMAT(1X,E15.8,E15.8)
6        FORMAT(1X,'HORIZONTAL LINE')
8        FORMAT(1X,1X,I3,1X,E15.8,1X,E15.8)
9        FORMAT(1X,'PROCEDURE COMPLETED SUCCESSFULLY')
11       FORMAT(1X,I3,4(1X,E15.8))
12       FORMAT(1X,'FAILURE')
        STOP
        END
$ENTRY
0430+1.00000000E-05
 6.00000000E+00 2.00000000E+01 5.00000000E+00-4.00000000E+01
+1.60000000E+01 0.50000000E+00-0.50000000E+00 0.00000000E+00
```

NEVILLE'S ITERATED INTERPOLATION ALGORITHM   3.1

```
C        TO EVALUATE THE INTERPOLATING POLYNOMIAL P ON THE
C        (N+1) DISTINCT NUMBERS X(0) ,..., X(N) AT THE NUMBER X
C        FOR THE FUNCTION F:
C
C        INPUT:   NUMBERS X(0) ,..., X(N) AS XX(1) ,..., XX(N+1);
C                 NUMBER X; VALUES OF F AS THE FIRST COLUMN OF Q.
```

156

```
C
C       OUTPUT:  THE TABLE Q WITH P(X) = Q(N+1,N+1).
C
C       DEFINE STORAGE FOR XX AND Q
        DIMENSION XX(10), Q(10,10), D(10)
C       INPUT:  DATA FOLLOWS THE END STATEMENT
        READ(5,1) N,X
        READ(5,2)(XX(I),I=1,N)
C       Q CAN BE GENERATED BY Q(I,1) = F(XX(I)) OR BE INPUT
        READ(5,2)(Q(I,1),I=1,N)
C       STEP 1
        D(1) = X-XX(1)
        DO 10 I=2,N
            D(I) = X-XX(I)
            DO 10 J=2,I
10      Q(I,J) = (D(I)*Q(I-1,J-1)-D(I-J+1)*Q(I,J-1))/(D(I)-D(I-J+1))
C       STEP 2
C       OUTPUT
        WRITE(6,3) X
        WRITE(6,4) N
        DO 20 I=1,N
20      WRITE(6,5) XX(I),(Q(I,J),J=1,I)
        WRITE(6,6)
        STOP
1       FORMAT(I2,1X,E15.8)
2       FORMAT(5(E15.8,1X))
3       FORMAT('1','TABLE FOR P EVALUATED AT X = ',E15.8,' FOLLOWS')
4       FORMAT('0',1X,'ENTRIES ARE XX(I),Q(I,1),...,Q(I,I) FOR EACH I
       *= 1 ,..., N WHERE N = ',I2)
5       FORMAT(('0',7(3X,E15.8)))
6       FORMAT('1')
        END
$ENTRY
 5             1.50E+00
        1.0E+00         1.3E+00         1.6E+00         1.9E+00
      2.2E+00     0.76519770E+00  0.62008600E+00  0.45540220E+00
    0.28181860E+00  0.11036230E+00
```

157

```
C      TO OBTAIN THE DIVIDED-DIFFERENCE COEFFICIENTS OF THE INTERPOLATORY
C      POLYNOMIAL P ON THE (N+1) DISTINCT NUMBERS X(0), X(1),..., X(N)
C      FOR THE FUNCTION F:
C
C      INPUT NUMBERS X(0),X(1),...,X(N); VALUES F(X(0)),F(X(1)),...,
C           F(X(N)) AS THE FIRST COLUMN Q(0,0),Q(1,0),...,Q(N,0) OF Q.
C
C      OUTPUT THE NUMBERS Q(0,0),Q(1,1),...,Q(N,N) WHERE
C           P(X) = Q(0,0)+Q(1,1)*(X-X(0))+Q(2,2)*(X-X(0))*(X-X(1))+
C                ...+Q(N,N)*(X-X(0))*(X-X(1))*...*(X-X(N-1)).
C
       DIMENSION X(5),Q(5,5)
       READ(5,1) N
C      THE INDICES ARE SHIFTED TO AVOID ZERO SUBSCRIPTS
       N=N+1
C      INPUT:   DATA FOLLOWS THE END STATEMENT
       READ(5,2) (X(I),Q(I,1),I=1,N)
       WRITE(6,3)
       WRITE(6,4) (X(I),Q(I,1),I=1,N)
C      STEP 1
       DO 10 I=2,N
          DO 20 J=2,I
20        Q(I,J)=(Q(I,J-1)-Q(I-1,J-1))/(X(I)-X(I-J+1))
10     CONTINUE
C      STEP 2
       WRITE(6,5) (Q(I,I),I=1,N)
       STOP
1      FORMAT(I2)
2      FORMAT(4(E15.8,1X))
3      FORMAT(1X,'INPUT DATA FOLLOWS')
4      FORMAT(1X,E15.8,1X,E15.8)
5      FORMAT(1X,'Q(0,0),...,Q(N,N)'/5(1X,E15.8))
       END
$ENTRY
04
 1.00000000E+00   0.76519770E+00   1.30000000E+00   0.62008600E+00
 1.60000000E+00   0.45540220E+00   1.90000000E+00   0.28181860E+00
 2.20000000E+00   0.11036230E+00
/*
```

```
C      TO OBTAIN THE COEFFICIENTS OF THE HERMITE INTERPOLATING
C      POLYNOMIAL H ON THE (N+1) DISTINCT NUMBERS X(0),...,X(N)
C      FOR THE FUNCTION F:
C
C      INPUT NUMBERS X(0), X(1),..., X(N); VALUES F(X(0)), F(X(1)),...,
C           F(X(N)) AND F'(X(0)), F'(X(1)),..., F'(X(N)).
C
C      OUTPUT NUMBERS Q(0,0), Q(1,1),..., Q(2N+1,2N+1) WHERE
C
C           H(X) = Q(0,0)+Q(1,1)*(X-X(0))+Q(2,2)*(X-X(0))**2+
C                  Q(3,3)*(X-X(0))**2*(X-X(1))+Q(4,4)*(X-X(0))**2*
C                  (X-X(1))**2+...+Q(2N+1,2N+1)*(X-X(0))**2*
C                  (X-X(1))**2*...*(X-X(N-1))**2*(X-X(N)).
C
C      DEFINE STORAGE FOR X, Z, Q
       DIMENSION X(3),Z(6),Q(6,6)
C      INPUT TO THE ALGORITHM: NOTE INPUT OF F(X(I)) AND F'(X(I)) ARE
C      INTO THE APPROPRIATE LOCATIONS IN Q

C      INPUT:   DATA FOLLOWS THE END STATEMENT
       READ(5,1) N
C      SHIFT INDICES TO AVOID ZERO SUBSCRIPTS
       N=N+1
       DO 10 I=1,N
10     READ(5,2) X(I),Q(2*I-1,1),Q(2*I,2)
C      STEP 1
       DO 20 I=1,N
C         STEP 2
          Z(2*I-1)=X(I)
          Z(2*I)=X(I)
          Q(2*I,1)=Q(2*I-1,1)
C         STEP 3
          IF(I.NE.1) THEN DO
             Q(2*I-1,2)=(Q(2*I-1,1)-Q(2*I-2,1))/(Z(2*I-1)-Z(2*I-2))
          ENDIF
20        CONTINUE
C      STEP 4
       K=2*N
       DO 30 I=3,K
          DO 30 J=3,I
30        Q(I,J)=(Q(I,J-1)-Q(I-1,J-1))/(Z(I)-Z(I-J+1))
```

```
C      STEP 5
       DO 40 I=1,K
40     WRITE(6,3) Z(I),(Q(I,J),J=1,I)
       STOP
1      FORMAT(I2)
2      FORMAT(3E15.8)
3      FORMAT(7(1X,E15.8))
       END
$ENTRY
 2
 1.30000000E+00 0.62008600E+00-0.52202320E+00
 1.60000000E+00 0.45540220E+00-0.56989590E+00
 1.90000000E+00 0.28181860E+00-0.58115710E+00
/*
```

NATURAL CUBIC SPLINE ALGORITHM   3.4

```
C      TO CONSTRUCT THE CUBIC SPLINE INTERPOLANT S FOR
C      THE FUNCTION F, DEFINED AT THE NUMBERS
C      X(0) < X(1) < ... < X(N), SATISFYING
C      S''(X(0)) = S''(X(N)) = 0:
C
C      INPUT:   N; X(O),X(1), ...,X(N); EITHER GENERATE
C               A(I) = F(X(I)) FOR I = 0,1, ...,N OR INPUT
C               A(I) FOR I = 0,1, ...,N.
C
C      OUTPUT:  A(J),B(J),C(J),D(J) FOR J = 0,1, ...,N-1.
C
C      NOTE:    S(X) = A(J) + B(J)*(X-X(J)) + C(J)*(X-X(J))**2 +
C                      D(J)*(X-X(J))**3 FOR X(J) < X < X(J+1)
C
       DIMENSION X(10),A(10),B(10),C(10),D(10),H(10),XA(10),XL(10),
      1XU(10),XZ(10)
C      INPUT:  DATA FOLLOWS THE END STATEMENT
       READ(5,1) N
C      THE SUBSCRIPTS ARE SHIFTED TO AVOID ZERO SUBSCRIPTS
       READ(5,2)(X(I),I=1,N)
       M = N-1
C      GENERATE A(I)=F(X(I)) AND H(I)=X(I+1)-X(I)
```

```
C      STEP 1
       DO 10 I=1,M
C            A(I)=F(X(I)), IF AVAILABLE
10     H(I) = X(I+1)-X(I)
C      A(N)=F(X(N)), IF AVAILABLE
       READ(5,2)(A(I),I=1,N)
C      STEP 2
       DO 20 I=2,M
C      USE XA INSTEAD OF ALPHA
20     XA(I) = 3*(A(I+1)*H(I-1)-A(I)*(X(I+1)-X(I-1))+A(I-1)*H(I))/
      *(H(I)*H(I-1))
C      STEP 3
C      STEPS 3,4,5 AND PART OF 6 SOLVE THE TRIDIAGONAL SYSTEM USING
C      ALGORITHM 6.7.
C
C      USE XL, XU, XZ IN PLACE OF L, MU, Z RESP.
       XL(1) = 1.0
       XU(1) = 0.0
       XZ(1) = 0.0
C      STEP 4
       DO 30 I=2,M
             XL(I) = 2*(X(I+1)-X(I-1))-H(I-1)*XU(I-1)
             XU(I) = H(I)/XL(I)
30     XZ(I) = (XA(I)-H(I-1)*XZ(I-1))/XL(I)
C      STEP 5
       XL(N) = 1.0
       XZ(N) = 0.0
       C(N) = XZ(N)
C      STEP 6
       DO 40 I=1,M
             J = N-I
             C(J) = XZ(J)-XU(J)*C(J+1)
             B(J) = (A(J+1)-A(J))/H(J)-H(J)*(C(J+1)+2*C(J))/3
40     D(J) = (C(J+1)-C(J))/(3*H(J))
C      STEP 7
C      OUTPUT
       WRITE(6,3) M
       WRITE(6,4)(X(I),A(I),B(I),C(I),D(I),I=1,M)
       WRITE(6,4) X(N)
       WRITE(6,5)
       STOP
1      FORMAT(I3)
2      FORMAT(5(E15.8,1X))
3      FORMAT('1','SPLINE CONSISTS OF',3X,I3,3X,'PIECES WITH COEF.
      * A(I),B(I),C(I),D(I)')
4      FORMAT(('0',5(E15.8,3X)))
5      FORMAT('1')
       END
```

$ENTRY
  4

27. 7E+00	28. 0E+00	29. 0E+00	30. 00E+00
4. 1E+00	4. 3E+00	4. 1E+00	3. 0E+00

CLAMPED CUBIC SPLINE ALGORITHM 3.5

```
C      TO CONSTRUCT THE CUBIC SPLINE INTERPOLANT S FOR
C      THE FUNCTION F, DEFINED AT THE NUMBERS
C      X(0) < X(1) < ... < X(N), SATISFYING
C      S'(X(0)) = F'(X(0)) AND S'(X(N)) = F'(X(N)):
C
C      INPUT:   N; X(0),X(1),...,X(N); EITHER GENERATE
C               A(I) = F(X(I)) FOR I = 0,1,...,N OR INPUT
C               A(I) FOR I = 0,1,...,N; FPO = F'(X(0)); FPN = F'(X(N)).
C
C      OUTPUT:  A(J),B(J),C(J),D(J) FOR J=0,1,...,N-1.
C
C      NOTE: S(X) = A(J) + B(J)*(X-X(J)) + C(J)*(X-X(J))**2 +
C                   D(J)*(X-X(J))**3 FOR X(J) < X < X(J+1)
C
       DIMENSION X(10),A(10),B(10),C(10),D(10),H(10),XA(10),XL(10),
      *XU(10),XZ(10)
C      INPUT: DATA FOLLOWS THE END STATEMENT
       READ(5,1) N
C      SUBSCRIPTS ARE SHIFTED TO AVOID ZERO SUBSCRIPTS
       READ(5,2)(X(I),I=1,N)
       M = N-1
C      GENERATE A(I)=F(X(I)) AND H(I)=X(I+1)-X(I)
C      STEP 1
       DO 10 I=1,M
C          A(I)=F(X(I)) IF AVAILABLE OR READ VALUES OF F
10     H(I) = X(I+1)-X(I)
C      A(N)=F(X(N)) IF AVAILABLE
       READ(5,2)(A(I),I=1,N)
       READ(5,2)  FPO, FPN
C      STEP 2
C      USE XA INSTEAD OF ALPHA
       XA(1) = 3*(A(2)-A(1))/H(1)-3*FPO
```

162

```fortran
      XA(N) = 3*FPN-3*(A(N)-A(N-1))/H(N-1)
C     STEP 3
      DO 20 I=2,M
20    XA(I) = 3*(A(I+1)*H(I-1)-A(I)*(X(I+1)-X(I-1))+A(I-1)*H(I))/
     *(H(I)*H(I-1))
C     STEP 4
C     STEPS 4,5,6 AND PART OF 7 SOLVE A TRIDIAGONAL LINEAR SYSTEM
C     USING ALGORITHM 6.7.
C
C     USE XL, XU, XZ IN PLACE OF L, MU, Z RESP.
      XL(1) = 2*H(1)
      XU(1) = 0.5
      XZ(1) = XA(1)/XL(1)
C     STEP 5
         DO 30 I=2,M
         XL(I) = 2*(X(I+1)-X(I-1))-H(I-1)*XU(I-1)
         XU(I) = H(I)/XL(I)
30    XZ(I) = (XA(I)-H(I-1)*XZ(I-1))/XL(I)
C     STEP 6
      XL(N) = H(N-1)*(2-XU(N-1))
      XZ(N) = (XA(N)-H(N-1)*XZ(N-1))/XL(N)
      C(N) = XZ(N)
C     STEP 7
      DO 40 I=1,M
         J = N-I
         C(J) = XZ(J)-XU(J)*C(J+1)
         B(J) = (A(J+1)-A(J))/H(J)-H(J)*(C(J+1)+2*C(J))/3
40    D(J) = (C(J+1)-C(J))/(3*H(J))
C     STEP 8
C     OUTPUT
      WRITE(6,3) M
      WRITE(6,4)(X(I),A(I),B(I),C(I),D(I),I=1,M)
      WRITE(6,4) X(N)
      WRITE(6,5)
      STOP
1     FORMAT(I3)
2     FORMAT(5(E15.8,1X))
3     FORMAT('1','SPLINE CONSISTS OF',3X,I3,3X,'PIECES WITH COEF.
     *A(I),B(I),C(I),D(I)')
4     FORMAT(('0',5(E15.8,3X)))
5     FORMAT('1')
      END
$ENTRY
   4
         27.7E+00          28.0E+00          29.0E+00          30.0E+00
          4.1E+00           4.3E+00           4.1E+00           3.0E+00
         0.33E+00          -1.5E+00
/*
```

```
C       TO APPROXIMATE I = INTEGRAL(( F(X)DX)) FROM A TO B:
C
C       INPUT:   ENDPOINTS A, B; EVEN POSITIVE INTEGER N.
C
C       OUTPUT:   APPROXIMATION XI TO I.
C
C       DEFINE FUNCTION F
        F(X) =SIN(X)
        PI=3.141593
C       INPUT:   DATA FOLLOWS THE END STATEMENT
        READ(5,1) A, B, N
C       STEP 1
        H = (B-A)/(2*N)
C       STEP 2
        XI0 = F(A) + F(B)
C       SUMMATION OF F(X(2*I-1))
        XI1 = 0.0
C       SUMMATION OF F(X(2*I))
        XI2 = 0.0
C       STEP 3
        MM=2*N-1
        DO 10 I=1,MM
C            STEP 4
             X = A+I*H
C            STEP 5
             IF (I.EQ.2*(I/2)) THEN DO
                  XI2 = XI2+F(X)
             ELSE DO
                  XI1 = XI1+F(X)
             END IF
10      CONTINUE
C       STEP 6
        XI = XI0+2*XI2+4*XI1
        XI = XI*H/3
C       STEP 7
C       OUTPUT
        WRITE(6,2) A, B, XI
        WRITE(6,3)
        STOP
1       FORMAT( 2E15.8, I3)
2       FORMAT('1', 'INTEGRAL OF F FROM', 3X, E15.8, 3X, 'TO', 3X, E15.8, 3X, 'IS'
```

```
      *, 3X, E15. 8)
3     FORMAT('1')
      END
$ENTRY
      0. 0E+00   3. 141593E+00    10
/*
```

```
C     TO APPROXIMATE THE I = INTEGRAL ((F(X)DX)) FORM A TO B TO WITHIN
C     A GIVEN TOLERANCE TOL > 0:
C
C     INPUT:    ENDPOINTS A, B; TOLERANCE TOL
C               LIMIT N TO NUMBER OF LEVELS
C
C     OUTPUT:   APPROXIMATION APP OR A MESSAGE THAT N IS EXCEED.
C
      DIMENSION TOL(20), A(20), H(20), FA(20), FC(20), FB(20), S(20), L(20)
      DIMENSION V(8)
      F(X)=  100/X**2*SIN(10/X)
C     USE AA, BB, EPS FOR A, B, TOL.
C     INPUT:  DATA FOLLOWS THE END STATEMENT
      READ(5, 1) AA, BB, EPS, N
C     STEP 1
      APP = 0
      I = 1
      TOL(I) = 10*EPS
      A(I) = AA
      H(I) = (BB-AA)/2
      FA(I) = F(AA)
      FC(I) = F(AA+H(I))
      FB(I) = F(BB)
C     APPROXIMATION FROM SIMPSON'S METHOD FOR ENTIRE INTERVAL
      S(I) = H(I)*(FA(I)+4*FC(I)+FB(I))/3
      L(I) = 1
C     STEP 2
      WHILE (I .GT. 0) DO
C           STEP 3
```

165

```
                 FD = F(A(I)+H(I)/2)
                 FE = F(A(I)+3*H(I)/2)
C                APPROXIMATIONS FROM SIMPSON'S METHOD FOR HALVES OF INTERVALS
                 S1 = H(I)*(FA(I)+4*FD+FC(I))/6
                 S2 = H(I)*(FC(I)+4*FE+FB(I))/6
C                SAVE DATA AT THIS LEVEL
                 V(1)=A(I)
                 V(2)=FA(I)
                 V(3)=FC(I)
                 V(4)=FB(I)
                 V(5)=H(I)
                 V(6)=TOL(I)
                 V(7)=S(I)
                 V(8)=L(I)
C                STEP 4
C                DELETE THE LEVEL
                 I=I-1
C                STEP 5
                 IF( ABS(S1+S2-V(7)) .LT. V(6)) THEN DO
                     APP = APP+(S1+S2)
                 ELSE DO
                     IF( V(8) .GE. N ) THEN DO
C                        PROCEDURE FAILS
                         WRITE(6,2)
                         STOP
                     ELSE DO
C                        ADD ONE LEVEL
C                        DATA FOR RIGHT HALF SUBINTERVAL
                         I = I+1
                         A(I) = V(1) + V(5)
                         FA(I) = V(3)
                         FC(I) = FE
                         FB(I) = V(4)
                         H(I) = V(5)/2
                         TOL(I) = V(6)/2
                         S(I) = S2
                         L(I) = V(8) + 1
C                        DATA FOR LEFT HALF SUBINTERVAL
                         I = I+1
                         A(I) = V(1)
                         FA(I) = V(2)
                         FC(I) = FD
                         FB(I) = V(3)
                         H(I) = H(I-1)
                         TOL(I) = TOL(I-1)
                         S(I) = S1
                         L(I) = L(I-1)
                     END IF
                 END IF
```

```
         END WHILE
C        STEP 6
C        OUTPUT
C        APP APPROXIMATES I TO WITHIN E
         WRITE(6,3) APP,EPS,AA,BB
         STOP
1        FORMAT(3E15.8,1X,I2)
2        FORMAT(1X,'LEVEL EXCEEDED')
3        FORMAT(3X,E15.8,' IS WITHIN ',E15.8,' FOR THE INTEGRAL FROM ',
        *E15.8,' TO ',E15.8)
         END
$ENTRY
+1.00000000E+00+3.00000000E+00+1.00000000E-04 20
/*
```

ROMBERG ALGORITHM   4.3

```
C        TO APPROXIMATE I = INTEGRAL((F(X)DX)) FROM A TO B:
C
C        INPUT:   ENDPOINTS A, B;  INTEGER N.
C
C        OUTPUT:   AN ARRAY R. ( R(2,N) IS THE APPROXIMATION TO I.)
C
C        R IS COMPUTED BY ROWS; ONLY 2 ROWS SAVED IN STORAGE
C
C        DEFINE STORAGE FOR TWO ROWS OF THE TABLE
         DIMENSION R(2,15)
C        DEFINE FUNCTION F
         F(X) = SIN(X)
C        INPUT:   DATA FOLLOWS THE END STATEMENT
         READ(5,1) N,A,B
C        STEP 1
         H = B-A
         R(1,1) = (F(A)+F(B))/2*H
C        STEP 2
         WRITE(6,3)
         WRITE(6,2) R(1,1)
C        STEP 3
         DO 10 I=2,N
```

```
C           STEP 4
C           APPROXIMATION FROM TRAPEZOIDAL METHOD
            SUM = 0.0
            M = 2**(I-2)
            DO 20 K=1,M
20          SUM = SUM+F(A+(K-.5)*H)
            R(2,1) = (R(1,1)+H*SUM)/2
C           STEP 5
C           EXTRAPOLATION
            DO 30 J=2,I
                 L = 2**(2*(J-1))
30          R(2,J) = (L*R(2,J-1)-R(1,J-1))/(L-1)
C           STEP 6
C           OUTPUT
            WRITE(6,2) (R(2,K),K=1,I)
C           STEP 7
            H = H/2
C           STEP 8
C           SINCE ONLY TWO ROWS ARE KEPT IN STORAGE, THIS STEP
C           IS TO PREPARE FOR THE NEXT ROW.
C           UPDATE ROW I OF R
            DO 10 J=1,I
10    R(1,J) = R(2,J)
C     STEP 9
      WRITE(6,3)
      STOP
1     FORMAT(I2,2(1X,E15.8))
2     FORMAT('0',(6(3X,E15.8)))
3     FORMAT('1')
      END
$ENTRY
 6 +0.00000000E+00 +3.14159300E+00
/*
```

```
C     TO APPROXIMATE I = DOUBLE INTEGRAL (( F(X,Y) DY DX )) WITH LIMITS
C         OF INTEGRATION FROM A TO B FOR X AND FROM C(X) TO D(X) FOR Y:
C
C     INPUT:    ENDPOINTS A,B; POSITIVE INTEGERS M, N.
C
C     OUTPUT:   APPROXIMATION J TO I.
C
C     LIMITS OF INTEGRATION
      C(X)=X**3
      D(X)=X**2
C     DEFINE FUNCTION F(X,Y)
      F(X,Y)=EXP(Y/X)
C     INPUT:  DATA FOLLOWS THE END STATEMENT
      READ(5,1) A,B,M,N
      NN=2*N+1
      MM=2*M-1
C     STEP 1
      H=(B-A)/(2*N)
C     USE AN, AE, AO FOR J(1), J(2), J(3) RESP.
C
C     END TERMS
      AN=0
C     EVEN TERMS
      AE=0
C     ODD TERMS
      AO=0
C     STEP 2
C     TO AVOID A ZERO SUBSCRIPT THE INDEX HAS BEEN SHIFTED BY ONE
      DO 10 I=1,NN
C        STEP 3
C        COMPOSITE SIMPSON'S METHOD FOR X
         X=A+(I-1)*H
         HX=(D(X)-C(X))/(2*M)
C        USE BN, BE, BO FOR K(1), K(2), K(3)
C
C        END TERMS
         BN=F(X,C(X))+F(X,D(X))
C        EVEN TERMS
         BE=0
```

```
C         ODD TERMS
          BO=0
C         STEP 4
          DO 20 J=1,MM
C            STEP 5
             Y=C(X)+J*HX
             Z=F(X,Y)
C            STEP 6
             IF(J.EQ.2*(J/2)) THEN DO
                 BE=BE+Z
             ELSE DO
                 BO=BO+Z
             END IF
20        CONTINUE
C      STEP 7
C      USE A1 FOR L, WHICH IS THE INTEGRAL OF F(X(I),Y) FROM C(X(I))
C         TO D(X(I)) BY COMPOSITE SIMPSON'S METHOD
       A1=(BN+2*BE+4*BO)*HX/3
C      STEP 8
       IF( I.EQ.1 .OR. I.EQ.NN ) THEN DO
          AN=AN+A1
       ELSE DO
          IF(I.EQ.2*(I/2)) THEN DO
              AO=AO+A1
          ELSE DO
              AE=AE+A1
          END IF
       END IF
10     CONTINUE
C      STEP 9
C      USE AC FOR J
       AC=(AN+2*AE+4*AO)*H/3
C      STEP 10
C      OUTPUT
       WRITE(6,2) AC
       STOP
1      FORMAT( 2E15.8, 1X, 2I2)
2      FORMAT( 1X, E15.8)
       END
$ENTRY
 0.10000000E+00 0.50000000E+00 0505
/*
```

```
C      TO APPROXIMATE I = TRIPLE INTEGRAL (( F(X,Y,Z) DZ DY DX )) WITH
C      LIMITS OF INTEGRATION FROM A TO B FOR X, FROM C(X) TO D(X) FOR Y
C      AND FROM ALPHA(X,Y) TO BETA(X,Y) FOR Z:
C
C      INPUT:    ENDPOINTS A,B; POSITIVE INTEGERS M, N, P.
C
C      OUTPUT:  APPROXIMATION J TO I.
C
C      LIMITS OF INTEGRATION
       INTEGER P,PP
       C(X)=-SQRT(4-X*X)
       D(X)=SQRT(4-X*X)
       ALPHA(X,Y) = SQRT(X*X+Y*Y)
       BETA(X,Y)=2
C      DEFINE FUNCTION F(X,Y)
       F(X,Y,Z)=SQRT(X*X+Y*Y)
C      INPUT:  DATA FOLLOWS THE END STATEMENT
       READ(5,1) A,B,M,N,P
       NN=2*N+1
       MM=2*M+1
       PP=2*P-1
C      STEP 1
       H=(B-A)/(2*N)
C      USE AN, AE, AO FOR J(1), J(2), J(3) RESP.
C
C      END TERMS
       AN=0
C      EVEN TERMS
       AE=0
C      ODD TERMS
       AO=0
C      STEP 2
C      TO AVOID A ZERO SUBSCRIPT THE INDEX HAS BEEN SHIFTED BY ONE
       DO 10 I=1,NN
C          STEP 3
C          COMPOSITE SIMPSON'S METHOD FOR FIXED X
           X=A+(I-1)*H
           HX=(D(X)-C(X))/(2*M)
C          USE BN, BE, BO FOR K(1), K(2), K(3) RESP.
C
C          END TERMS
```

171

```
              BN=0
C             EVEN TERMS
              BE=0
C             ODD TERMS
              BO=0
C             STEP 4
              DO 20 J=1,MM
C                 STEP 5
C                 COMPOSITE SIMPSON'S METHOD FOR FIXED X AND Y
                  Y=C(X)+(J-1)*HX
                  ZA=ALPHA(X,Y)
                  ZB=BETA(X,Y)
                  HY=(ZB-ZA)/(2*P)
C                 USE CN, CE, CO FOR L(1), L(2), L(3) RESP.
C                 END TERMS
                  CN = F(X,Y,ZA)+F(X,Y,ZB)
C                 EVEN TERMS
                  CE=0
C                 ODD TERMS
                  CO=0
C                 STEP 6
                  DO 30 K=1,PP
C                     STEP 7
                      Z=ZA+K*HY
                      Q=F(X,Y,Z)
C                     STEP 8
                      IF(K.EQ.2*(K/2)) THEN
                          CE = CE+Q
                      ELSE
                          CO=CO+Q
                      ENDIF
30                CONTINUE
C                 STEP 9
C                 USE B1 FOR L
                  B1 = (CN+2*CE+4*CO)*HY/3
C                 STEP 10
                  IF (J.EQ.1 .OR. J.EQ.MM) THEN
                      BN=BN+B1
                  ELSE
                      IF(J.EQ.2*(J/2)) THEN DO
                          BO=BO+B1
                      ELSE DO
                          BE=BE+B1
                      END IF
                  ENDIF
20            CONTINUE
C         STEP 11
C         USE A1 FOR L
          A1=(BN+2*BE+4*BO)*HX/3
```

```
C       STEP 12
        IF( I.EQ.1 .OR. I.EQ.NN ) THEN DO
           AN=AN+A1
        ELSE DO
           IF(I.EQ.2*(I/2)) THEN DO
              AO=AO+A1
           ELSE DO
              AE=AE+A1
           END IF
        END IF
10      CONTINUE
C       STEP 13
C       USE AC FOR J
        AC=(AN+2*AE+4*AO)*H/3
C       STEP 14
C       OUTPUT
        WRITE(6,2) AC
        STOP
1       FORMAT( 2E15.8, 1X, 3I2)
2       FORMAT( 1X, E15.8)
        END
$ENTRY
-2.00000000E+00 2.00000000E+00 080808
/*
```

EULER'S ALGORITHM   5.1

```
C       TO APPROXIMATE THE SOLUTION THE INITIAL VALUE PROBLEM
C               Y' = F(T,Y), A <= T <= B, Y(A) = ALPHA,
C       AT N+1 EQUALLY SPACED POINTS IN THE CLOSED INTERVAL A,B.
C
C       INPUT:    ENDPOINTS A,B;  INTEGER N;  INITIAL CONDITION ALPHA.
C
C       OUTPUT:   APPROXIMATION W TO Y AT THE (N+1) VALUES OF T.
C
        F(T,W)=-W+T+1
C       INPUT:  DATA FOLLOWS THE END STATEMENT
        READ(5,2) A, B, N, ALPHA
C       STEP 1
        H=(B-A)/N
```

173

```
          T=A
          W=ALPHA
          WRITE(6,1) T,W
C         STEP 2
          DO 10 I =1,N
C              STEP 3
C              COMPUTE W(I)
               W=W+H*F(T,W)
C              COMPUTE T(I)
               T=A+I*H
C              STEP 4
10             WRITE(6,1) T,W
C         STEP 5
          STOP
1         FORMAT(3X,'FOR T= ',E15.8,3X,'W= ',E15.8)
2         FORMAT(2E15.8,I3,E15.8)
          END
$ENTRY
 0.00000000E+00 1.00000000E+00 10 1.00000000E+00
/*
```

RUNGE-KUTTA (ORDER FOUR) ALGORITHM   5.2

```
C     TO APPROXIMATE THE SOLUTION TO THE INITIAL VALUE PROBLEM
C             Y' = F(T,Y), A <= T <= B, Y(A) = ALPHA,
C     AT (N+1) EQUALLY SPACED NUMBERS IN THE INTERVAL (A,B).
C
C     INPUT:   ENDPOINTS A,B; INTEGER N; INITIAL CONDITION ALPHA.
C
C     OUTPUT:  APPROXIMATION W TO Y AT THE (N+1) VALUES OF T.
C
      F(T,W)=-W+T+1
C     INPUT:  DATA FOLLOWS THE END STATEMENT
      READ(5,1) A,B,N,ALPHA
C     STEP 1
      H = (B-A)/N
      T=A
      W=ALPHA
      WRITE(6,2)
      I=0
```

174

```
        WRITE(6,3) I,T,W
C       STEP 2
        DO 10 I=1,N
C           STEP 3
C           USE XK1,XK2,XK3,XK4 FOR K(1),K(2),K(3),K(4) RESP.
            XK1 = H*F(T,W)
            XK2 = H*F(T+H/2,W+XK1/2)

            XK3 = H*F(T+H/2,W+XK2/2)
            XK4 = H*F(T+H,W+XK3)
C           STEP 4
C           COMPUTE W(I)
            W = W + (XK1+2*(XK2+XK3)+XK4)/6
C           COMPUTE T(I)
            T=A+I*H
C           STEP 5
10          WRITE(6,3) I, T, W
        WRITE(6,4)
C       STEP 6
        STOP
1       FORMAT(2E15.8,I3,E15.8)
2       FORMAT('1',7X,'I',12X,'T',19X,'W')
3       FORMAT('0',5X,I3,5X,E15.8,5X,E15.8)
4       FORMAT('1')
        END
$ENTRY
 0.00000000E+00  1.00000000E+00  10  1.00000000E+00
/*
```

RUNGE-KUTTA-FEHLBERG ALGORITHM   5.3

```
C       TO APPROXIMATE THE SOLUTION OF THE INITIAL VALUE PROBLEM
C                   Y'=F(T,Y), A<=T<=B, Y(A)=ALPHA,
C       WITH LOCAL TRUNCATION ERROR WITHIN A GIVEN TOLERANCE.
C
C       INPUT:     ENDPOINTS A,B; INITIAL CONDITION ALPHA; TOLERANCE
C                  TOL; MAXIMUM STEPSIZE HMAX; MINIMUM STEPSIZE HMIN.
C
C       OUTPUT:    T,W,H WHERE W APPROXIMATES Y(T) AND STEPSIZE H
C                  IS USED OR A MESSAGE THAT MINIMUM STEPSIZE WAS EXCEEDED.
C
```

```
C        INITIALIZATION AND INPUT
         F(T,W)=-W+T+1
C        INPUT:   DATA FOLLOWS THE END STATEMENT
         READ(5,5) A,B,TOL
         READ(5,5) ALPHA,HMAX,HMIN
         WRITE(6,1)
         WRITE(6,2)
C        STEP 1
         H=HMAX
         W=ALPHA
         T=A
         WRITE(6,3) T,W
         IFLAG = 1
         IFLAG1=0
C        STEP 2
         WHILE((T.LE.B) .AND. (IFLAG.EQ.1)) DO
C             STEP 3
              XK1=H*F(T,W)
              XK2=H*F(T+H/4,W+XK1/4)
              XK3=H*F(T+3*H/8,W+(3*XK1+9*XK2)/32)
              XK4=H*F(T+12*H/13,W+(1932*XK1-7200*XK2+7296*XK3)/2197)

              XK5=H*F(T+H,W+439*XK1/216-8*XK2+3680*XK3/513-845*XK4/4104)
              XK6=H*F(T+H/2,W-8*XK1/27+2*XK2-3544*XK3/2565+1859*XK4/4104
     *        -11*XK5/40)
C             STEP 4
              R=ABS(XK1/360-128*XK3/4275-2197*XK4/75240+XK5/50+2*XK6/55)/H
C             STEP 5
C             TO AVOID UNDERFLOW
              IF(R.GT.1.0E-20) THEN DO
                   DELTA=.84*(TOL/R)**.25
              ELSE DO
                   DELTA=10.0
              END IF
C             STEP 6
              IF(R.LE.TOL) THEN DO
C             STEP 7
C             APPROXIMATION ACCEPTED
                   T = T + H
                   W=W+25*XK1/216+1408*XK3/2565+2197*XK4/4104-XK5/5
C             STEP 8
                   WRITE(6,3) T,W,H,R
                   IF(IFLAG1.EQ.1) IFLAG=0
              END IF
C             STEP 9
C             CALCULATE NEW H
              IF(DELTA.LE..1) THEN DO
                   H = .1*H
              ELSE DO
                   IF(DELTA.GE.4.) THEN DO
```

176

```
                              H = 4*H
                ELSE DO
                        H = DELTA*H
                END IF
        END IF
C       STEP 10
        IF(H.GT.HMAX) H=HMAX
C       STEP 11
        IF(H.LT.HMIN) THEN DO
                IFLAG=0
                WRITE(6,4)
        ELSE DO
                IF (T+H.GT.B) THEN DO
                        IF (ABS(B-T).LT.TOL) THEN DO
                                T=B
                        ELSE DO
                                H=B-T
                        ENDIF
                        IFLAG1=1
                ENDIF
        END IF
    END WHILE
C       STEP 12
C       THE PROCESS IS COMPLETE.
    WRITE(6,1)
    STOP
1       FORMAT('1')
2       FORMAT(1X,'ORDER OF OUTPUT IS: T,W(I),H,R',/)
3       FORMAT(1X,4(E15.8,1X))
4       FORMAT(1X,'MINIMAL H EXCEEDED')
5       FORMAT(3E15.8)
    END

$ENTRY
  0.00000000E+00 2.00000000E+00 1.00000000E-05
  1.00000000E+00 2.50000000E-01 2.00000000E-02
/*
```

```
C      TO APPROXIMATE THE SOLUTION OF THE INITIAL VALUE PROLEM:
C               Y'=F(T,Y), A<=T<=B, Y(A)=ALPHA,
C      AT (N+1) EQUALLY SPACED NUMBERS IN THE INTERVAL (A,B).
C
C      INPUT:   ENDPOINTS A,B; INTEGER N; INITIAL CONDITION ALPHA.
C
C      OUTPUT:  APPROXIMATION W TO Y AT THE (N+1) VALUES OF T.
C
C      T(1),...,T(4) AND W(1),...,W(4) ARE THE 4 MOST RECENT VALUES OF
C      T(I) AND W(I) RESP.
C
       DIMENSION T(4),W(4)
       F(T,W) = -W+T+1
C      INPUT:   DATA FOLLOWS THE END STATEMENT
       READ(5,1) A,B,N,ALPHA
C      STEP 1
C      THE SUBSCRIPTS ARE SHIFTED TO AVOID ZERO SUBSCRIPTS
       T(1) = A
       W(1) = ALPHA
       H = (B-A)/N
       I = 0
       WRITE(6,2)
       WRITE(6,3) I,T(1),W(1)
C      STEP 2
       DO 10 I=1,3
C           STEPS 3 AND 4
C           COMPUTE STARTING VALUES USING RUNGE-KUTTA METHOD GIVEN IN A
C           SUBROUTINE—NOTE:  FUNCTION F IS NEEDED IN THE SUBROUTINE
            CALL RK4(H,T(I),W(I),T(I+1),W(I+1))
C           STEP 5
10     WRITE(6,3) I,T(I+1),W(I+1)
C      STEP 6
       DO 20 I=4,N
C           STEP 7
C           TO, WO WILL BE USED IN PLACE OF T, W RESP.
            TO=A+I*H
C           PREDICT W(I)
            WO = W(4)+H*(55*F(T(4),W(4))-59*F(T(3),W(3))+37*F(T(2),W(2)
     *      )-9*F(T(1),W(1)))/24
C           CORRECT W(I)
            WO = W(4)+H*(9*F(TO,WO)+19*F(T(4),W(4))-5*F(T(3),W(3))+
     *      F(T(2),W(2)))/24
```

```
C          STEP 8
           WRITE(6,3) I,TO,WO
C          STEP 9
C          PREPARE FOR NEXT ITERATION
           DO 30 J=1,3
               T(J) = T(J+1)
30             W(J) = W(J+1)
C          STEP 10
           T(4) = TO
           W(4) = WO
20     CONTINUE
       WRITE(6,4)
C      STEP 11
       STOP
1      FORMAT(2E15.8,I3,E15.8)
2      FORMAT('1',7X,'I',12X,'T',19X,'W')
3      FORMAT('0',5X,I3,5X,E15.8,5X,E15.8)
4      FORMAT('1')
       END
           SUBROUTINE RK4(H,TO,WO,TI,WI)
           F(T,W) = -W+T+1
           TI = TO+H
           XK1 = H*F(TO,WO)
           XK2 = H*F(TO+H/2,WO+XK1/2)
           XK3 = H*F(TO+H/2,WO+XK2/2)
           XK4 = H*F(TI,WO+XK3)
           WI = WO+(XK1+2*(XK2+XK3)+XK4)/6
           RETURN
           END
$ENTRY
 0.00000000E+00 1.00000000E+00 10 1.00000000E+00
/*
```

```
C       TO APPROXIMATE THE SOLUTION OF THE INITIAL VALUE PROBLEM
C                 Y'=F(T,Y), A <= T <= B , Y(A)=ALPHA,
C       WITH LOCAL TRUNCATION ERROR WITHIN A GIVEN TOLERANCE:
C
C       INPUT:    ENDPOINTS A,B; INITIAL CONDITION ALPHA; TOLERANCE
C                 TOL; MAXIMUM STEPSIZE HMAX; MINIMUM STEPSIZE HMIN.
C
C       OUTPUT:   I,T(I),W(I),H WHERE AT THE ITH STEP W(I) APPROXIMATES
C                 Y(T(I)) AND STEPSIZE H WAS USED OR A MESSAGE THAT THE
C                 MINIMUM STEP SIZE WAS EXCEEDED.
C
C       STEP 1 SETS UP THE SUBALGORITHM FOR RK4
C       THE FUNCTION F MUST ALSO BE DEFINED IN THE SUBROUTINE
C
        INTEGER FLAG
        DIMENSION T(100),W(100)
        F(T,W)= -W+T+1
C       INPUT:  DATA FOLLOWS THE END STATEMENT
        READ(5,5) A,B,ALPHA
        READ(5,5) TOL,HMAX,HMIN
C       STEP 2
C       SUBSCRIPTS ARE SHIFTED TO AVOID ZERO SUBSCRIPTS
        T(1)= A
        W(1)= ALPHA

        H= HMAX
C       FLAG WILL BE USED TO EXIT THE LOOP IN STEP 4
        FLAG=1
C       LAST WILL INDICATE WHEN THE LAST VALUE IS CALCULATED.
        LAST=0
        WRITE(6,1) T(1),W(1)
C       STEP 3
        KK=1
        CALL RK4(KK,H,W,T)
C       NFLAG INDICATES COMPUTATION FROM RK4
        NFLAG=1
        I=5
C       USE TT IN PLACE OF T
        TT=T(4)+H
C       STEP 4
        WHILE(FLAG .EQ. 1 ) DO
C           STEP 5
```

```
C          PREDICT W(I)
           WP=W(I-1)+H*(55*F(T(I-1),W(I-1))-59*F(T(I-2),W(I-2))+37*F(T(I-
      1    3),W(I-3))-9*F(T(I-4),W(I-4)))/24
C          CORRECT W(I)
           WC=W(I-1)+H*(9*F(TT,WP)+19*F(T(I-1),W(I-1))-5*F(T(I-2),W(I-2))
      1    +F(T(I-3),W(I-3)))/24
           SIG=19*(ABS(WC-WP))/(270*H)
C          STEP 6
           IF(SIG.LE.TOL) THEN DO
C               STEP 7
C               RESULT ACCEPTED
                W(I)=WC
                T(I)=TT
C               STEP 8
                IF(NFLAG.EQ.1) THEN DO
                     K=I-3
                     KK=I-1
                     DO 10 J=K,KK
 10                  WRITE(6,2) J,T(J),W(J),H
                     WRITE(6,2) I,T(I),W(I),H,SIG
                ELSE DO
                     WRITE(6,3) I,T(I),W(I),H,SIG
                END IF
C               STEP 9
                IF (LAST.EQ.1) THEN DO
                     FLAG=0
                ELSE DO
C                    STEP 10
                     I=I+1
                     NFLAG=0
C                    STEP 11
                     IF((SIG.LE.0.1*TOL).OR.(T(I-1)+H.GT.B)) THEN DO
C                    INCREASE H IF IT MORE ACCURATE THAN REQUIRED OR
C                    DECREASE H TO INCLUDE B AS A MESH POINT.
C                    STEP 12
C                    TO AVOID UNDERFLOW
                          IF (SIG .LE. 1.0E-70) THEN DO
                               Q=4
                          ELSE DO
                               Q=(TOL/(2*SIG))**.25
                          END IF
C                         STEP 13
                          IF(Q.GT.4) THEN DO
                               H=4*H
                          ELSE DO
                               H=Q*H
                          END IF
C                         STEP 14
                          IF(H.GT.HMAX) H=HMAX
```

181

```
C                         STEP 15
C                    AVOID TERMINATING WITH CHANGE IN STEPSIZE
                          IF(T(I-1)+4*H.GE.B) THEN DO
                              H=(B-T(I-1))/4
                              LAST=1
                          ENDIF
C                         STEP 16
                          NFLAG=1
                          KK=I-1
                          CALL RK4(KK,H,W,T)
                          I=I+3
                      END IF
                  END IF
              ELSE DO
C             RESULT REJECTED
C             STEP 17
              Q=(TOL/(2*SIG))**.25
C             STEP 18
              IF(Q.LT.0.1) THEN DO
                      H=0.1*H
              ELSE DO
                  H=Q*H
              END IF
C             STEP 19
              IF(H.LT.HMIN) THEN DO
C                     PROCEDURE FAILS
                      WRITE(6,4)
                      FLAG=0
              ELSE DO
C             PREVIOUS RESULTS ALSO REJECTED
                      IF(NFLAG.EQ.1) I=I-3
                      KK=I-1
                      CALL RK4(KK,H,W,T)
                      I=I+3
                      NFLAG=1
              ENDIF
          END IF
C     STEP 20
      TT=T(I-1)+H
      END WHILE
C     STEP 21
      STOP
1     FORMAT(1X,'T(1)=',E15.8,5X,'W(1)=',E15.8)
2     FORMAT(1X,'J=',I3,1X,'T=',E15.8,1X,'W=',E15.8,1X,'H=',4(1X,E15.8))
3     FORMAT(1X,'I=',I3,1X,'T=',E15.8,1X,'W=',E15.8,1X,'H=',4(1X,E15.8))
4     FORMAT(1X,'HMIN EXCEEDED - PROCEDURE FAILS')
5     FORMAT(3E15.8)
      END
      SUBROUTINE RK4(K,H,V,X)
```

```
      DIMENSION V(100),X(100)
      F(X,V)=-V+X+1
      DO 20 I=1,3
      J=K+I-1
      XK1=H*F(X(J),V(J))
      XK2=H*F(X(J)+H/2,V(J)+XK1/2)
      XK3=H*F(X(J)+H/2,V(J)+XK2/2)
      XK4=H*F(X(J)+H,V(J)+XK3)
      V(J+1)=V(J)+(XK1+2*(XK2+XK3)+XK4)/6
20    X(J+1)=X(J)+H
      RETURN
      END
$ENTRY
  0.00000000E+00  2.00000000E+00  1.00000000E+00
  1.00000000E-05  2.50000000E-01  2.00000000E-02
/*
```

EXTRAPOLATION ALGORITHM   5.6

```
C       TO APPROXIMTE THE SOLUTION OF THE INITIAL VALUE PROBLEM:
C                 Y'=F(T,Y),  A<=T<=B,  Y(A)=ALPHA,
C       WITH LOCAL ERROR WITHIN A GIVEN TOLERANCE:
C
C       INPUT:    ENDPOINTS A,B;  INITIAL CONDITION ALPHA;
C                 TOLERANCE TOL;  MAXIMUM STEPSIZE HMAX;
C                 MINIMUM STEPSIZE HIMIN.
C
C       OUTPUT:   T,W,H WHERE W APPROXIMATES Y(T) AND STEPSIZE H WAS
C                 USED OR A MESSAGE THAT MINIMUM STEPSIZE EXCEEDED.
C
      DIMENSION NK(8),Y(8),Q(7,7)
C       STEP 1
      DATA NK/2,3,4,6,8,12,16,24/
      F(T,W)= -W+T+1
C       INPUT:  DATA FOLLOWS THE END STATEMENT
      READ(5,3) A,B,ALPHA
      READ(5,3) TOL,HMAX,HMIN
C       STEP 2
      TO=A
```

```
        WO=ALPHA
        H=HMAX
C       IFLAG IS USED TO EXIT THE LOOP IN STEP 4
        IFLAG=1
C       STEP 3
        DO 10 I=1,7
        DO 10 J=1,I
C       ( Q(I,J) = H(J)**2/H(I+1)**2  )
10      Q(I,J)=(FLOAT(NK(I+1))/FLOAT(NK(J)))**2
C       STEP 4
        WHILE(IFLAG.EQ.1)DO
C           STEP 5
            K=1
C           WHEN DESIRED ACCURACY ACHIEVED, NFLAG IS SET TO 1
            NFLAG=0
C           STEP 6
            WHILE(K.LE.8 .AND. NFLAG.EQ.0) DO
C               STEP 7
                HK=H/NK(K)
                T=TO
                W2=WO
C               EULER FIRST STEP
                W3=W2+HK*F(T,W2)
                T=TO+HK
C               STEP 8
                M=NK(K)-1
                DO 20 J=1,M
                    W1=W2
                    W2=W3
C                   MIDPOINT METHOD
                    W3=W1+2*HK*F(T,W2)
20              T=TO+(J+1)*HK
C               STEP 9
C               SMOOTHING TO COMPUTE Y(K,1)
                Y(K)=(W3+W2+HK*F(T,W3))/2
C               STEP 10
C   NOTE: Y(K-1)=Y(K-1,1), Y(K-2)=Y(K-1,2),..., Y(1)=Y(K-1,K-1) SINCE
C       ONLY PREVIOUS ROW OF TABLE REQUIRED
                IF(K.GE.2) THEN DO
C                   STEP 11
                    J=K
C                   SAVE Y(K-1,K-1)
                    V=Y(1)
C                   STEP 12
                    WHILE(J.GE.2) DO
C                   EXTRAPOLATION TO COMPUTE Y(J-1)=Y(K,K-J+2)
C   NOTE: Y(J-1) = (H(J-1)**2*Y(J)-H(K)**2*Y(J-1))/(H(J-1)**2-H(K)**2)
                        Y(J-1)=Y(J)+(Y(J)-Y(J-1))/(Q(K-1,J-1)-1)
                        J=J-1
```

```
                                END WHILE
C                               STEP 13
                                IF(ABS(Y(1)-V).LE.TOL) NFLAG=1
C                               Y(1) ACCEPTED AS NEW W
                        END IF
C                       STEP 14
                        K=K+1
                END WHILE
C           STEP 15
            K=K-1
C           STEP 16
            IF(NFLAG.EQ.0) THEN DO
C                   STEP 17
C                   NEW VALUE FOR W REJECTED, DECREASE H
                    H=H/2
C                   STEP 18
                    IF(H.LT.HMIN) THEN DO
                            WRITE(6,1)
                            IFLAG=0
                    END IF
            ELSE DO
C                   STEP 19
C                   NEW VALUE FOR W ACCEPTED
                    WO=Y(1)
                    TO=TO+H
C                   STEP 20
                    WRITE(6,2) TO,WO,K,H
                    IF(ABS(TO-B).LT.HMIN/2)THEN DO
                        IFLAG=0
                    ELSE DO
                        IF(TO+H.GT.B)THEN DO
                            H=B-TO
                        ELSE DO
                            IF(K.LE.3.AND.H.LT.HMAX/2) THEN DO
                                H=2*H
                            ENDIF
                        ENDIF
                    ENDIF
            END IF
        END WHILE
C       STEP 21
        STOP
1       FORMAT(1X,'FAILURE')
2       FORMAT(1X,'Y(',E15.8,') IS APPROX. BY ',E15.8,': OBTAINED WITH K='
     *   ,I3,' AND H = ',E15.8)
3       FORMAT(3E15.8)
        END
$ENTRY
 0.00000000E+00 2.00000000E+00 1.00000000E+00
```

185

```
1.00000000E-05 2.50000000E-01 2.00000000E-02
/*
```

```
C       TO APPROXIMATE THE SOLUTION OF THE MTH-ORDER SYSTEM OF FIRST-
C       ORDER INITIAL-VALUE PROBLEMS
C               UJ' = FJ(T,U1,U2,...,UM), J=1,2,...,M
C           A <= T <= B,  UJ(A)=ALPHAJ, J=1,2,...,M
C       AT (N+1) EQUALLY SPACED NUMBERS IN THE INTERVAL A<=T<=B.
C
C       INPUT ENDPOINTS A,B; NUMBER OF EQUATIONS M; INTEGER N; INTIAL
C           CONDITIONS ALPHA1,...,ALPHAM.
C
C       OUTPUT APPROXIMATIONS WJ TO UJ(T) AT THE (N+1) VALUES OF T.
C
C       DEFINE FUNCTIONS F1,...,FM
        F1(T,X1,X2) = -4*X1+3*X2+6
        F2(T,X1,X2) = -2.4*X1+1.6*X2+3.6
C       INPUT:   DATA FOLLOWS THE END STATEMENT
        READ(5,2) A,B,N
        READ(5,2) ALPHA1,ALPHA2
C       STEP 1
        H=(B-A)/N
        T=A
C       STEP 2
        W1=ALPHA1
        W2=ALPHA2
C       STEP 3
        WRITE(6,1) T,W1,W2
C       STEP 4
        DO 10 I=1,N
C          STEP 5
           X11=H*F1(T,W1,W2)
           X12=H*F2(T,W1,W2)
C          STEP 6
           X21=H*F1(T+H/2,W1+X11/2,W2+X12/2)
           X22=H*F2(T+H/2,W1+X11/2,W2+X12/2)
C          STEP 7
           X31=H*F1(T+H/2,W1+X21/2,W2+X22/2)
           X32=H*F2(T+H/2,W1+X21/2,W2+X22/2)
```

```
C          STEP 8
           X41=H*F1(T+H,W1+X31,W2+X32)
           X42=H*F2(T+H,W1+X31,W2+X32)
C          STEP 9
           W1=W1+(X11+2*X21+2*X31+X41)/6
           W2=W2+(X12+2*X22+2*X32+X42)/6
C          STEP 10
           T=A+I*H
C          STEP 11
           WRITE(6,1) T,W1,W2
10         CONTINUE
C     STEP 12
      STOP
1     FORMAT(3(5X,E15.8))
2     FORMAT(2E15.8,I3)
      END
$ENTRY
  0.00000000E+00  0.50000000E+00   5
  0.00000000E+00  0.00000000E+00
/*
```

TRAPEZOIDAL WITH NEWTON ITERATION ALGORITHM   5.8

```
C     TO APPROXIMATE THE SOLUTION OF THE INITIAL-VALUE PROBLEM
C              Y' = F(T,Y), A <= T <= B, Y(A)=ALPHA
C     AT (N+1) EQUALLY SPACED NUMBERS IN THE INTERVAL A<=T<=B:
C
C     INPUT ENDPOINTS A,B; INTEGER N, INITIAL CONDITION ALPHA;
C          TOLERANCE TOL; MAXIMUM NUMBER OF ITERATIONS M AT ANY ONE STEP.
C
C     OUTPUT APPROXIMATION W TO Y AT THE (N+1) VALUES OF T OR A MESSAGE
C          OF FAILURE.
C
C     DEFINE FUNCTION F AND ITS PARTIAL WITH RESPECT TO Y
C
      F(T,W)=-6*W+6
      FYP(T,W)=-6
C     INPUT:  DATA AFTER THE END STATEMENT
      READ(5,3) A,B,N
      READ(5,3) ALPHA,TOL,M
```

```
C       STEP 1
        W=ALPHA
        T=A
        H=(B-A)/N
        WRITE(6,1) T,W
C       STEP 2
        DO 10 I=1,N
C          STEP 3
           XK1=W+H/2*F(T,W)
           WO=XK1
           J=1
           IFLAG=0
C          STEP 4
           WHILE ( IFLAG.EQ.0) DO
C              STEP 5
               W=WO-(WO-XK1-H/2*F(T+H,WO))/(1-H/2*FYP(T+H,WO))
C              STEP 6
               IF(ABS(W-WO).LT.TOL) THEN DO
                   IFLAG=1
               ELSE DO
                   J=J+1
                   WO=W
                   IF(J.GT.M) THEN DO
                       WRITE(6,2)
                       STOP
                     ENDIF
                 ENDIF
           END WHILE
C          STEP 7
           T=A+I*H
           WRITE(6,1) T,W
10      CONTINUE
C       STEP 8
        STOP
1       FORMAT(1X,2(E15.8,1X))
2       FORMAT(1X,'METHOD FAILS')
3       FORMAT(2E15.8,I3)
        END
$ENTRY
 0.00000000E+00 1.00000000E+00 20
 2.00000000E+00 1.00000000E-04 10
/*
```

```
C     TO SOLVE THE N BY N LINEAR SYSTEM
C
C  E1:  A(1,1) X(1) + A(1,2) X(2) + ... + A(1,N) X(N) = A(1,N+1)
C  E2:  A(2,1) X(1) + A(2,2) X(2) + ... + A(2,N) X(N) = A(2,N+1)
C  :
C  .
C  EN:  A(N,1) X(1) + A(N,2) X(2) + ... + A(N,N) X(N) = A(N,N+1)
C
C   INPUT:   NUMBER OF UNKNOWNS AND EQUATIONS N; AUGMENTED
C            MATRIX A = (A(I,J)) WHERE 1<=I<=N AND 1<=J<=N+1.
C
C   OUTPUT:  SOLUTION X(1),X(2),...,X(N) OR A MESSAGE THAT THE LINEAR
C            SYSTEM HAS NO UNIQUE SOLUTION.
C
C   INITIALZATION
      DIMENSION A(4,5), X(4)
C   INPUT:  DATA FOLLOWS THE END STATEMENT
      READ(5,1) N
      M = N+1
      READ(5,2)((A(I,J),J=1,M),I=1,N)
C  ICHG COUNTS NUMBER OF INTERCHANGES
      ICHG = 0
      WRITE(6,3)
      WRITE(6,4)((A(I,J),J=1,M),I=1,N)
C   STEP 1
C   ELIMINATION PROCESS
      NN = N-1
      DO 10 I=1,NN
C         STEP 2
C         USE IP IN PLACE OF P
          IP = I
          WHILE(ABS(A(IP,I)).LT.1.0E-20 .AND. IP.LE.N) DO
              IP = IP+1
          END WHILE
          IF(IP.EQ.N+1)THEN DO
              SYSTEM DOES NOT HAVE UNIQUE SOLUTION
              WRITE(6,5) ((A(I,J),J=1,M),II=1,N)
              STOP
          END IF
C         STEP 3
          IF(IP.NE.I) THEN DO
              DO 20 JJ=1,M
```

189

```
                              C = A(I,JJ)
                              A(I,JJ) = A(IP,JJ)
20                    A(IP,JJ) = C
                      ICHG = ICHG+1
              END IF
C             STEP 4
              JJ = I+1
              DO 30 J=JJ,N
C                     STEP 5
C                     NOT SAVING MULTIPLIERS IN THIS ALGORITHM
C                     USE XM IN PLACE OF M(J,I)
                      XM = A(J,I)/A(I,I)
C                     STEP 6
                      DO 40 K=JJ,M
40                        A(J,K) = A(J,K)-XM*A(I,K)
30            A(J,I) = 0
10      CONTINUE
C       STEP 7
        IF(ABS(A(N,N)).LT.1.0E-20) THEN DO
C               SYSTEM DOES NOT HAVE UNIQUE SOLUTION
                WRITE(6,5)((A(I,J),J=1,M),I=1,N)
                STOP
        END IF
C       STEP 8
C       START BACKWARD SUBSTITUTION
        X(N) = A(N,N+1)/A(N,N)
C       STEP 9
        L = N-1
        DO 15 K=1,L
              I = L-K+1
              JJ = I+1
              SUM = A(I,N+1)
              DO 16 KK=JJ,N
16            SUM = SUM-A(I,KK)*X(KK)
15      X(I) = SUM/A(I,I)
        WRITE(6,6)((A(I,J),J=1,M),I=1,N)
C       STEP 10
C       PROCEDURE COMPLETED SUCCESSFULLY
        WRITE(6,7)(X(I),I=1,N)
        WRITE(6,8) ICHG
        STOP
1       FORMAT(I2)
2       FORMAT(5E15.8)
5       FORMAT('1','THE PRECEDING SYSTEM HAS NO UNIQUE SOLUTION')
4       FORMAT(1X,5(3X,E15.8))
6       FORMAT('0','THE REDUCED SYSTEM: '/(5(3X,E15.8)))
7       FORMAT('0','HAS SOLUTION VECTOR',4(3X,E15.8))

8       FORMAT('0','NUMBER OF INTERCHANGES = ',3X,I2)
3       FORMAT('1','ORIGINAL SYSTEM',/)
```

```
        END
$ENTRY
  4
   1.00000000E+00-1.00000000E+00 2.00000000E+00-1.00000000E+00
 -8.00000000E+00 2.00000000E+00-2.00000000E+00 3.00000000E+00
 -3.00000000E+00-2.00000000E+01 1.00000000E+00 1.00000000E+00
   1.00000000E+00 0.00000000E+00-2.00000000E+00 1.00000000E+00
 -1.00000000E+00 4.00000000E+00 3.00000000E+00 4.00000000E+00
```

GAUSSIAN MAXIMAL COLUMN PIVOTING ALGORITHM 6.2

```
C       TO SOLVE A N BY N LINEAR SYSTEM
C
C   E1:  A(1,1) X(1) + A(1,2) X(2) + ... + A(1,N) X(N) = A(1,N+1)
C   E2:  A(2,1) X(1) + A(2,2) X(2) + ... + A(2,N) X(N) = A(2,N+1)
C   :
C   .
C   EN:  A(N,1) X(1) + A(N,2) X(2) + ... + A(N,N) X(N) = A(N,N+1)
C
C    (MULTIPLIERS WILL BE SAVED AND PRINTED AS PART OF REDUCED
C     SYSTEM):
C
C     INPUT:    NUMBER OF UNKNOWNS AND EQUATIONS N; AUGMENTED MATRIX
C               A=(A(I,J)) WHERE 1<=I<=N AND 1<=J<=N+1.
C
C     OUTPUT:   SOLUTION X(1),X(2),...,X(N) OR A MESSAGE THAT THE
C               LINEAR SYSTEM HAS NO UNIQUE SOLUTION.
C
C     INITIALIZATION
      DIMENSION A(4,5),NROW(4)
C     INPUT:  DATA FOLLOWS THE END STATEMENT
      READ(5,1) N
      M=N+1
      READ(5,2) ((A(I,J),J=1,M),I=1,N)
C     ICHG COUNTS NUMBER OF INTERCHANGES
      ICHG=0
      WRITE(6,3)
      WRITE(6,4) ((A(I,J),J=1,M),I=1,N)
C     STEP 1
```

```
         DO 10 I=1,N
C     INITIALIZE ROW POINTER
10       NROW(I)=I
C     STEP 2
         NN = N-1
         DO 20 I=1,NN
C          STEP 3
           IMAX=NROW(I)
           AMAX=ABS(A(IMAX, I))
           IMAX=I
           JJ=I+1
           DO 30 IP=JJ,N
               JP=NROW(IP)
               IF(ABS(A(JP, I)).GT.AMAX) THEN DO
                   AMAX=ABS(A(JP, I))
                   IMAX=IP
               END IF
30         CONTINUE
C          STEP 4
           IF(AMAX.LT.1.0E-20) THEN DO
               WRITE(6,5)
               STOP
           END IF
C          STEP 5
           IF(NROW(I).NE.NROW(IMAX)) THEN DO
C          SIMULATE ROW INTERCHANGE
               ICHG = ICHG+1
               NCOPY=NROW(I)
               NROW(I)=NROW(IMAX)
               NROW(IMAX)=NCOPY
           END IF
C          STEP 6
           I1=NROW(I)
           DO 40 J=JJ,N
               J1=NROW(J)
C              STEP 7
               XM=A(J1, I)/A(I1, I)
C              STEP 8
               DO 50 K=JJ,M
50             A(J1, K)=A(J1, K)-XM*A(I1, K)
40         A(J1, I)=XM
20       CONTINUE
C     STEP 9
         N1=NROW(N)
         IF(ABS(A(N1,N)).LT.1.0E-20) THEN DO
C          SYSTEM HAS NO UNIQUE SOLUTION
               WRITE(6,5)
               WRITE(6,11)
               STOP
```

192

```
          END IF
C         STEP 10
C         START BACKWARD SUBSTITUTION
          A(N1,N+1)=A(N1,N+1)/A(N1,N)
C         STEP 11
          DO 60 K=1,NN
                I=NN-K+1
                JJ=I+1
                N2=NROW(I)
                SUM=A(N2,N+1)
                DO 70 KK=JJ,N
                     LL = NROW(KK)
70              SUM=SUM-A(N2,KK)*A(LL,N+1)
60        A(N2,N+1)=SUM/A(N2,I)
C         STEP 12
C         PROCEDURE COMPLETED SUCCESSFULLY
          WRITE(6,6) ((A(I,J),J=1,N),I=1,N)
          WRITE(6,7) (A(NROW(I),M),I=1,N)
          WRITE(6,8) ICHG
          WRITE(6,9) (NROW(I),I=1,N)
          WRITE(6,11)
          STOP
1         FORMAT(I2)
2         FORMAT(5(E15.8,1X))
5         FORMAT('0','THE PRECEDING SYSTEM HAS NO UNIQUE SOLUTION')
4         FORMAT(('0',5(3X,E15.8)))
6         FORMAT('0',/1X,'THE REDUCED SYSTEM:  '//(/,4(3X,E15.8)))
7         FORMAT('0','HAS SOLUTION VECTOR',4(3X,E15.8))
8         FORMAT('0','NUMBER OF INTERCHANGES = ',3X,I2)
3         FORMAT('1',1X,'ORIGINAL SYSTEM:',/)
9         FORMAT(1X,4I5)
11        FORMAT('1')
          END
$ENTRY
 4
          1.000000        -1.000000        2.000000       -1.000000
      -8.000000    2.000000       -2.000000        3.000000       -3.000000
     -20.000000    1.000000        1.000000        1.000000        0.000000
      -2.000000    1.000000       -1.000000        4.000000        3.000000
       4.000000
```

```
C      TO SOLVE AN N BY N LINEAR SYSTEM
C
C
C   E1:   A(1,1) X(1) + A(1,2) X(2) + ... + A(1,N) X(N) = A(1,N+1)
C   E2:   A(2,1) X(1) + A(2,2) X(2) + ... + A(2,N) X(N) = A(2,N+1)
C   :
C   .
C   EN:   A(N,1) X(1) + A(N,2) X(2) + ... + A(N,N) X(N) = A(N,N+1)
C
C     INPUT:    NUMBER OF UNKNOWNS AND EQUATIONS N; AUGMENTED MATRIX
C               A=(A(I,J)) WHERE 1<=I<=N AND 1<=J<=N+1.
C
C     OUTPUT:   SOLUTION X(1),X(2),...X(N) OR A MESSAGE THAT LINEAR
C               SYSTEM HAS NO UNIQUE SOLUTION.
C
C     INITIALIZATION
      DIMENSION A(4,5),NROW(4),S(4)
C     INPUT:   DATA FOLLOWS THE END STATEMENT
      READ(5,1) N
      M=N+1
      READ(5,2) ((A(I,J),J=1,M),I=1,N)
C     ICHG COUNTS NUMBER OF INTERCHANGES
      ICHG=0
      WRITE(6,3)
      WRITE(6,4) ((A(I,J),J=1,M),I=1,N)
C     STEP 1
      DO 10 I=1,N
C     INITIALIZE ROW POINTER
10    NROW(I)=I
      DO 20 I=1,N
          S(I)=ABS(A(I,1))
          DO 30 J=1,N
              IF(ABS(A(I,J)).GT.S(I)) S(I)=ABS(A(I,J))
30        CONTINUE
          IF(S(I).LT.1.0E-20) THEN DO
C             SYSTEM HAS NO UNIQUE SOLUTION
              WRITE(6,5)
              STOP
          END IF
20    CONTINUE
C     STEP 2
C     ELIMINATION PROCESS
```

194

```
          NN = N-1
          DO 90 I=1,NN
C             STEP 3
              IMAX=NROW(I)
              AMAX=ABS(A(IMAX,I))/S(IMAX)
              IMAX=I
              JJ=I+1
              DO 40 IP=JJ,N
                  JP=NROW(IP)
                  IF(ABS(A(JP,I))/S(JP).GT.AMAX) THEN DO
                      AMAX=ABS(A(JP,I))/S(JP)
                      IMAX=IP
                  END IF
40            CONTINUE
C             STEP 4
              IF(AMAX.LT.1.0E-20) THEN DO
C                 SYSTEM HAS NO UNIQUE SOLUTION
                  WRITE(6,5)
                  STOP
              END IF
C             STEP 5
C             SIMULATE ROW INTERCHANGE
              IF(NROW(I).NE.NROW(IMAX)) THEN DO
                  ICHG = ICHG+1
                  NCOPY=NROW(I)
                  NROW(I)=NROW(IMAX)
                  NROW(IMAX)=NCOPY
              END IF
C             STEP 6
              I1=NROW(I)
              DO 50 J=JJ,N
                  J1=NROW(J)
C                 STEP 7
                  XM=A(J1,I)/A(I1,I)
C                 STEP 8
                  DO 60 K=JJ,M
60                A(J1,K)=A(J1,K)-XM*A(I1,K)
50            A(J1,I)=XM
90        CONTINUE
C     STEP 9
      N1=NROW(N)
      IF(ABS(A(N1,N)).LT.1.0E-20) THEN DO
C         SYSTEM HAS NO UNIQUE SOLUTION
          WRITE(6,5)
          WRITE(6,11)
          STOP
      END IF
C     STEP 10
C     START BACKWARD SUBSTITUTION
```

```
C          STORE SOLUTION IN (N+1)ST COLUMN OF A
           A(N1,N+1)=A(N1,N+1)/A(N1,N)
C          STEP 11
           DO 70 K=1,NN
                I=NN-K+1
                JJ=I+1
                N2=NROW(I)
                SUM=A(N2,N+1)
                DO 80 KK=JJ,N
                     LL = NROW(KK)
80              SUM=SUM-A(N2,KK)*A(LL,N+1)
70         A(N2,N+1)=SUM/A(N2,I)
C          STEP 12
C          PROCEDURE COMPLETED SUCCESSFULLY
           WRITE(6,6) ((A(I,J),J=1,N),I=1,N)
           WRITE(6,7) (A(NROW(I),M),I=1,N)
           WRITE(6,8) ICHG
           WRITE(6,9) (NROW(I),I=1,N)
           WRITE(6,11)
           STOP
1          FORMAT(I2)
2          FORMAT(5(E15.8,1X))
5          FORMAT('0','THE PRECEDING SYSTEM HAS NO UNIQUE SOLUTION')
4          FORMAT(('0',5(3X,E15.8)))
6          FORMAT('0',/1X,'THE REDUCED SYSTEM: '//(/,4(3X,E15.8)))
7          FORMAT('0','HAS SOLUTION VECTOR',4(3X,E15.8))
8          FORMAT('0','NUMBER OF INTERCHANGES = ',3X,I2)
3          FORMAT('1',1X,'ORIGINAL SYSTEM: ',/)
9          FORMAT(1X,4I5)
11         FORMAT('1')
           END
$ENTRY
  4
          1.000000       -1.000000        2.000000       -1.000000
     -8.000000    2.000000        -2.000000        3.000000        -3.000000
     -20.000000    1.000000        1.000000        1.000000        0.000000
     -2.000000    1.000000        -1.000000        4.000000        3.000000
      4.000000
```

```
C     TO FACTOR THE N BY N MATRIX A=(A(I,J)) INTO THE PRODUCT OF THE
C     LOWER TRIANGULAR MATRIX L= L(I,J)  AND U= U(I,J) , THAT IS A=LU,
C     WHERE THE MAIN DIAGONAL OF EITHER L OR U IS GIVEN:
C
C     INPUT:     DIMENSION N; THE ENTRIES A(I,J), 1<=I, J<=N, OF A; THE
C                DIAGONAL L(1,1),...L(N,N) OF L OR THE DIAGONAL
C                U(1,1),...U(N,N) OF U.
C
C     OUTPUT:    THE ENTRIES L(I,J), 1<=J<=I, 1<=I<=N OF L AND THE
C                ENTRIES U(I,J), I<=J<=N, 1<=I<=N OF U.
C
C     INITIALIZATION
      DIMENSION A(4,4),XL(4)
C     INPUT:  DATA FOLLOWS THE END STATEMENT
      READ(5,1) N
      WRITE(6,2)
      WRITE(6,3)
      READ(5,4) ((A(I,J),J=1,N),I=1,N)
C
C     ISW = 0 IF DIAGONAL OF L INPUT; ISW = 1 IF DIAGONAL
C     OF U INPUT.
C
      READ(5,9) ISW
      DO 80 I=1,N
80    XL(I) = 1.0
      WRITE(6,5)((A(I,J),J=1,N),I=1,N)
C     STEP 1
      IF ( ABS(A(1,1)) .LT. 1.0E-20 ) THEN DO
           WRITE(6,8)
           STOP
      END IF
C     THE ENTRIES OF L BELOW THE MAIN DIAGONAL WILL BE PLACED IN THE
C     CORRESPONDING ENTRIES OF A; THE ENTRIES OF U ABOVE THE MAIN
C     DIAGONAL WILL BE PLACED IN THE CORRESPONDING ENTRIES OF A;
C     THE MAIN DIAGONAL WHICH WAS NOT INPUT WILL BECOME THE MAIN
C     DIAGONAL OF A; THE INPUT MAIN DIAGONAL OF L OR U IS, OF COURSE,
C     PLACED IN XL.
      A(1,1)=A(1,1)/XL(1)
C     STEP 2
      DO 10 J=2,N
      IF(ISW.EQ.0) THEN DO
C           FIRST ROW OF U
```

197

```
                A(1,J)=A(1,J)/XL(1)
C               FIRST COLUMN OF L
                A(J,1)=A(J,1)/A(1,1)
       ELSE DO
C               FIRST ROW OF U
                A(1,J)=A(1,J)/A(1,1)
C               FIRST COLUMN OF L
                A(J,1)=A(J,1)/XL(1)
       END IF
10     CONTINUE
C      STEP 3
       M=N-1
       DO 20 I=2,M
C           STEP 4
            KK=I-1
            S=0
            DO 30 K=1,KK
30          S=S-A(I,K)*A(K,I)
            A(I,I)=(A(I,I)+S)/XL(I)
            IF(ABS(A(I,I)).LT.1.0E-20) THEN DO
                 WRITE(6,8)
                 STOP
            END IF
C           STEP 5
            JJ=I+1
            DO 40 J=JJ,N
                 SS=0
                 S=0
                 DO 50 K=1,KK
                      SS=SS-A(I,K)*A(K,J)
50                    S=S-A(J,K)*A(K,I)
                 IF(ISW.EQ.0) THEN DO
C                     ITH ROW OF U
                      A(I,J)=(A(I,J)+SS)/XL(I)
C                     ITH COLUMN OF L
                      A(J,I)=(A(J,I)+S)/A(I,I)
                 ELSE DO
C                     ITH ROW OF U
                      A(I,J)=(A(I,J)+SS)/A(I,I)
C                     ITH COLUMN OF L
                      A(J,I)=(A(J,I)+S)/XL(I)
                 END IF
40          CONTINUE
20     CONTINUE
C      STEP 6
       S=0
       DO 60 K=1,M
60     S=S-A(N,K)*A(K,N)
       A(N,N)=(A(N,N)+S)/XL(N)
```

```
C        IF A(N,N) = 0 THEN A = LU BUT THE MATRIX IS SINGULAR
C        PROCESS IS COMPLETE, ALL ENTRIES OF A HAVE BEEN DETERMINED
C        STEP 7
         WRITE(6,6)
         WRITE(6,11) ISW
         WRITE(6,5) (XL(I),I=1,N)
         WRITE(6,7)
         WRITE(6,5) ((A(I,J),J=1,N),I=1,N)
         STOP
1        FORMAT(I2)
2        FORMAT('1')
3        FORMAT(1X,'ORIGINAL MATRIX A')
4        FORMAT(4F10.5)
5        FORMAT(4(5X,E15.8))
6        FORMAT(1X,'DIAGONAL OF L OR U')
7        FORMAT(1X,'ENTRIES OF L BELOW/ON DIAGONAL AND ENTRIES OF U ABOVE/
        *ON DIAGONAL')
8        FORMAT('0',' FACTORIZATION IMPOSSIBLE')
9        FORMAT(I1)
11       FORMAT('0','IF 0 THEN L AND IF 1 THEN U, IT IS ',I2)
         END
$ENTRY
 4
6.0        2.0        1.0        -1.0
2.0        4.0        1.0         0.0
1.0        1.0        4.0        -1.0
-1.0       0.0        -1.0        3.0
0
/*
```

LDL^t ALGORITHM   6.5

```
C        TO FACTOR THE POSITIVE DEFINITE N BY N MATRIX A INTO LDL**T,
C        WHERE L IS LOWER TRIANGULAR WITH ONES ALONG THE DIAGONAL AND D
C        IS A DIAGONAL MATRIX WITH POSITIVE ENTRIES ON THE DIAGONAL
C
C        INPUT:    THE DIMENSION N; ENTRIES A(I,J), 1<=I, J<=N OF A.
C
C        OUTPUT:   THE ENTRIES L(I,J), 1<=J<I, 1<=I<=N OF L AND D(I),
C                  1<=I<=N OF D.
```

```
C       INITIALIZATION
        DIMENSION A(3,3),D(3),V(3)
C       INPUT:   DATA FOLLOWS THE END STATEMENT
        READ(5,1) N
        READ(5,2) ((A(I,J),J=1,N),I=1,N)
        WRITE(6,3)
        WRITE(6,4)
        WRITE(6,5) ((A(I,J),J=1,N),I=1,N)
C       STEP 1
        DO 50 I = 1, N
C          STEP 2
           II = I-1
           IF (I .GT. 1 ) THEN
                DO 10 J=1,II
10              V(J) = A(I,J)*D(J)
           ENDIF
C          STEP 3
           D(I) = A(I,I)
           IF (I .GT. 1 ) THEN
                DO 20 J=1,II
20              D(I) = D(I)-A(I,J)*V(J)
           ENDIF
C          STEP 4
           JJ = I+1
           IF (I .LT. N) THEN
                DO 30 J=JJ,N
                    IF (I .GT. 1) THEN
                        DO 40 K=1,II
40                      A(J,I)=A(J,I)-A(J,K)*V(K)
                    ENDIF
30              A(J,I)=A(J,I)/D(I)
           ENDIF
50      CONTINUE
C       STEP 5
        WRITE(6,6)
        DO 70 I=2,N
           II=I-1
70      WRITE(6,5) (A(I,J),J=1,II)
        WRITE(6,7)
        WRITE(6,5) (D(I),I=1,N)
        WRITE(6,3)
        STOP
1       FORMAT(I2)
2       FORMAT(3F10.5)
3       FORMAT('1')
4       FORMAT(1X,'ORIGINAL MATRIX A')
5       FORMAT(1X,3(5X,E15.8))
6       FORMAT(1X,'MATRIX L')
7       FORMAT(1X,'DIAGONAL D')
```

```
          END
$ENTRY
  3
 4.0         -1.0          1.0
 -1.0        4.25          2.75
  1.0        2.75          3.5
```

CHOLESKI ALGORITHM   6.6

```
C     TO FACTOR THE POSITIVE DEFINITE N BY N MATRIX A INTO LL**T,
C     WHERE L IS LOWER TRIANGULAR.
C
C     INPUT:    THE DIMENSION N; ENTRIES A(I,J), 1<=I, J<=N OF A.
C
C     OUTPUT:   THE ENTRIES L(I,J), 1<=J<=I, 1<=I<=N OF L.
C
C     THE ENTRIES OF U=L**T ARE U(I,J)=L(J,I), I<=J<=N, 1<=I<=N
C
C     INITIALIZATION
      DIMENSION A(3,3)
C     INPUT:   DATA FOLLOWS THE END STATEMENT
      READ(5,1) N
      READ(5,2) ((A(I,J),J=1,N),I=1,N)
      WRITE(6,3)
      WRITE(6,4)
      WRITE(6,5) ((A(I,J),J=1,N),I=1,N)
C     STEP 1
      A(1,1)=SQRT(A(1,1))
C     STEP 2
      DO 10 J=2,N
10    A(J,1)=A(J,1)/A(1,1)
C     STEP 3
      NN=N-1
      DO 20 I=2,NN
C        STEP 4
         KK=I-1
         S=0
C        DO-LOOP COMPUTES THE SUMMATION
         DO 30 K=1,KK
30          S=S-A(I,K)**2
         A(I,I)=SQRT(A(I,I)+S)
```

201

```
C              STEP 5
               JJ=I+1
               DO 40 J=JJ,N
                      S=0
                      KK=I-1
C                     DO-LOOP COMPUTES THE SUMMATION
                      DO 50 K=1,KK
50                         S=S-A(J,K)*A(I,K)
40                    A(J,I)=(A(J,I)+S)/A(I,I)
20             CONTINUE
C       STEP 6
        S=0
        DO 60 K=1,NN
60          S=S-A(N,K)**2
        A(N,N)=SQRT(A(N,N)+S)
C       STEP 7
        WRITE(6,6)
        DO 70 I=1,N
70      WRITE(6,5) (A(I,J),J=1,I)
        WRITE(6,3)
        STOP
1       FORMAT(I2)
2       FORMAT(3F10.5)
3       FORMAT('1')
4       FORMAT(1X,'ORIGINAL MATRIX A')
5       FORMAT(1X,3(5X,E15.8))
6       FORMAT(1X,'MATRIX L')
        END
$ENTRY
 3
4.0         -1.0          1.0
-1.0         4.25         2.75
 1.0         2.75         3.5
```

```
C      TO SOLVE AN N BY N LINEAR SYSTEM
C
C      E1:   A(1,1) X(1) + A(1,2) X(2)                          = A(1,N+1)
C      E2:   A(2,1) X(1) + A(2,2) X(2) + A(2,3) X(3)            = A(2,N+1)
C      :
C      .
C      EN:                     A(N-1,N) X(N-1) + A(N,N) X(N)    = A(N,N+1)
C
C
C      INPUT:    THE DIMENSION N; THE ENTRIES OF A.
C
C      OUTPUT:   THE SOLUTION X(1),..,X(N).
C
C      INITIALIZATION
       DIMENSION A(4),B(4),C(4),BB(4),Z(4),X(4)
C      INPUT:   DATA FOLLOWS THE END STATEMENT
       READ(5,2) N
C      A(I,I) IS STORED IN A(I),  1<=I<=N
       READ(5,3) (A(I),I=1,N)
C      THE LOWER SUB-DIAGONAL A(I,I-1) IS STORED IN B(I), 2<=I<=N
       READ(5,3) (B(I),I=2,N)
C      THE UPPER SUB-DIAGONAL A(I,I+1) IS STORED IN C(I),  1<=I<=N-1
       NN=N-1
       READ(5,3) (C(I),I=1,NN)
C      A(I,N+1) IS STORED IN BB(I),  1<=I<=N
       READ(5,3) (BB(I),I=1,N)
       WRITE(6,1)
       WRITE(6,4)
       WRITE(6,5) A(1),C(1),BB(1)
       WRITE(6,6) ( B(I),I-1,A(I),I,C(I),I+1,BB(I) ,I=2,NN)
       WRITE(6,7) B(N),A(N),BB(N)
C      STEP 1
C      THE ENTRIES OF U OVERWRITE C AND THE ENTRIES OF L OVERWRITE A
       C(1)=C(1)/A(1)
C      STEP 2
       DO 10 I=2,NN
C           ITH ROW OF L
            A(I)=A(I)-B(I)*C(I-1)
C           (I+1)ST COLUMN OF U
10          C(I)=C(I)/A(I)
C      STEP 3
C      NTH ROW OF L
```

```fortran
          A(N)=A(N)-B(N)*C(N-1)
C         STEP 4
C         STEPS 4,5 SOLVE LZ = B
          Z(1)=BB(1)/A(1)
C         STEP 5
          DO 20 I=2,N
20            Z(I)=(BB(I)-B(I)*Z(I-1))/A(I)
C         STEP 6
C         STEPS 6,7 SOLVE UX = Z
          X(N)=Z(N)
C         STEP 7
          DO 30 II=1,NN
              I=NN-II+1
30            X(I)=Z(I)-C(I)*X(I+1)
C         STEP 8
          WRITE(6,8)
          WRITE(6,9) (X(I),I=1,N)
          WRITE(6,1)
          STOP
1         FORMAT('1')
2         FORMAT(I2)
3         FORMAT(5F10.5)
4         FORMAT(1X,'ORIGINAL SYSTEM')
5         FORMAT(1X,E15.8,' X( 1) + ',E15.8,' X( 2) ',25X,'= ',E15.8)
6         FORMAT(1X,E15.8,' X(',I2,') + ',E15.8,' X(',I2,') + ',E15.8,' X('
     *,I2,') = ',E15.8)
7         FORMAT(1X,23X,E15.8,' X(N-1) + ',E15.8,' X( N) = ',E15.8)
8         FORMAT(1X,'THE SOLUTION VECTOR IS')
9         FORMAT(1X,4(5X,E15.8))
          END
$ENTRY
 4
2.0        2.0        2.0        2.0
-1.0       -1.0       -1.0
-1.0       -1.0       -1.0
1.0        0.0        0.0        1.0
```

```
C       JACOBI ITERATIVE ALGORITHM    7.1
```

```
C       TO SOLVE AX = B GIVEN AN INITIAL APPROXIMATION X(0).
C
C       INPUT:    THE NUMBER OF EQUATIONS AND UNKNOWNS N; THE ENTRIES
C                 A(I,J), 1<=I, J<=N, OF THE MATRIX A; THE ENTRIES B(I),
C                 1<=I<=N, OF THE INHOMOGENEOUS TERM B; THE ENTRIES
C                 XO(I), 1<=I<=N, OF X(0); TOLERANCE TOL; MAXIMUM
C                 NUMBER OF ITERATIONS N.
C
C       OUTPUT:   THE APPROXIMATE SOLUTION X(1),...,X(N) OR A MESSAGE

C                 THAT THE NUMBER OF ITERATIONS WAS EXCEEDED.
C
C       INITIALIZATION.
        DIMENSION A(4,5),X1(4),X2(4)
C       INPUT:  DATA FOLLOWS THE END STATEMENT
C       USE NN FOR CAPITAL N
        READ(5,1) N, TOL, NN
        M = N+1
C       USE A(I,N+1) = B(I) FOR 1<=I<=N
        READ(5,2)((A(I,J),J=1,M),I=1,N)
C       USE X1 FOR XO
        READ(5,2)(X1(I),I=1,N)
        WRITE(6,3)
        WRITE(6,4)((A(I,J),J=1,M),I=1,N)
        WRITE(6,5)
        WRITE(6,4)(X1(I),I=1,N)
C       STEP 1
        K = 1
C       STEP 2
        WHILE(K.LE.NN)DO
C            ERR IS USED TO TEST ACCURACY-IT MEASURES THE L2 NORM
             ERR = 0.0
C            STEP 3
             DO 10 I=1,N
                  S = 0.0
C                 DO-LOOP COMPUTED THE SUMMATION
                  DO 20 J=1,N
20                     S = S-A(I,J)*X1(J)
                  S = (S+A(I,N+1))/A(I,I)
                  ERR = ERR+S*S
C                 USE X2 FOR X
10                X2(I) = X1(I)+S
```

```
                ERR = SQRT(ERR)
                WRITE(6,6) K,ERR,(X2(I),I=1,N)
C               STEP 4
                IF(ERR.LE.TOL) THEN DO
C                    PROCESS IS COMPLETE
                    WRITE(6,7) K
                    WRITE(6,8)
                    STOP
                END IF
C               STEP 5
                K = K+1
C               STEP 6
                DO 30 I=1,N
30                  X1(I) = X2(I)
        END WHILE
C       STEP 7
C       PROCEDURE COMPLETED UNSUCCESSFULLY
        WRITE(6,9) NN
        STOP
1       FORMAT(I2,E15.8,I3)
2       FORMAT(5(E15.8,1X))
3       FORMAT('1','ORIGINAL SYSTEM: '/)
4       FORMAT(('0',5(3X,E15.8)))
5       FORMAT('0','INITIAL APPROXIMATION: '/)
6       FORMAT('0','ITERATION NUMBER',I3,' GIVES ERROR ',E15.8,' FOR APPROX
     *. ',4(2X,E15.8))
7       FORMAT('0','CONVERGENCE ON INTERATION NUMBER ',I4)
8       FORMAT('1')
9       FORMAT('0','FAILURE AFTER INTERATION NUMBER ',I4)
        END
$ENTRY
 4 1.00000000E-03 30
            1.0E+01          -1.0E+00          2.0E+00          .000E+00
            6.0E+00          -1.0E+00          1.1E+01          -1.0E+00
            3.0E+00           2.5E+01          2.0E+00          -1.0E+00
            1.0E+01          -1.0E+00         -1.1E+01           0.0E+00
            3.0E+00          -1.0E+00          8.0E+00           1.5E+01
            0.0E+00           0.0E+00          0.0E+00           0.0E+00
```

```
C       TO SOLVE AX = B GIVEN AN INITIAL APPROXIMATION X(0):
C
C       INPUT:    THE NUMBER OF EQUATIONS AND UNKNOWNS N; THE ENTRIES
C                 A(I,J), 1<=I, J<=N, OF THE MATRIX A; THE ENTRIES B(I)
C                 1<=I<=N, OF THE INHOMOGENEOUS TERM B; THE ENTRIES
C                 XO(I), 1<=I<=N, OF X(0); TOLERANCE TOL; MAXIMUM
C                 NUMBER OF ITERATIONS N.
C
C       OUTPUT:   THE APPROXIMATE SOLUTION X(1),...,X(N) OR A MESSAGE
C                 THAT THE NUMBER OF ITERATIONS WAS EXCEEDED.
C
C       INITIALIZATION.
        DIMENSION A(4,5),X1(4)
C       INPUT:   DATA FOLLOWS THE END STATEMENT
C       USE NN FOR CAPITAL N
        READ(5,1) N,TOL,NN
        M = N+1
C       B(I) = A(I,N+1) FOR 1<=I<=N
        READ(5,2)((A(I,J),J=1,M),I=1,N)
C       USE X1 FOR XO
        READ(5,2)(X1(I),I=1,N)
        WRITE(6,3)
        WRITE(6,4)((A(I,J),J=1,M),I=1,N)
        WRITE(6,5)
        WRITE(6,4)(X1(I),I=1,N)
C       STEP 1
        K = 1
C       STEP 2
        WHILE(K.LE.NN)DO
C          ERR IS USED TO TEST ACCURACY AND MEASURES THE L2 NORM
           ERR = 0.0
C          STEP 3
           DO 10 I=1,N
                S = 0.0
C               DO-LOOP COMPUTED THE SUMMATION
                DO 20 J=1,N
20                   S = S-A(I,J)*X1(J)
                S = (S+A(I,N+1))/A(I,I)
                ERR = ERR+S*S
10              X1(I) = X1(I)+S
           ERR = SQRT(ERR)
```

```
                  WRITE(6,6) K,ERR,(X1(I),I=1,N)
C                 STEP 4
                  IF(ERR.LE.TOL) THEN DO
C                      PROCESS IS COMPLETE
                       WRITE(6,7) K
                       WRITE(6,8)
                       STOP
                  END IF
C                 STEP 5
                  K = K+1
                  IF(K.GT.NN)THEN DO
                       WRITE(6,9) NN
                       WRITE(6,8)
                       STOP
                  END IF
C     STEP 6—IS NOT USED SINCE ONLY ONE VECTOR IS REQUIRED
      END WHILE
C     STEP 7
C     PROCEDURE COMPLETED UNSUCCESSFULLY
      WRITE(6,9) NN
      STOP
1     FORMAT(I2,E15.8,I3)
2     FORMAT(5(E15.8,1X))
3     FORMAT('1','ORIGINAL SYSTEM: '/)
4     FORMAT(('0',5(3X,E15.8)))
5     FORMAT('0','INITIAL APPROXIMATION:'/)
6     FORMAT('0','ITERATION NUMBER',I3,' GIVES ERROR ',E15.8,' FOR APPROX
     *.  ',4(2X,E15.8))
7     FORMAT('0','CONVERGENCE ON ITERATION NUMBER ',I4)
8     FORMAT('1')
9     FORMAT('0','FAILURE AFTER ITERATION NUMBER ',I4)
      END
$ENTRY
 4 1.00000000E-03 30
          1.0E+01        -1.0E+00         2.0E+00          .000E+00
          6.0E+00        -1.0E+00         1.1E+01         -1.0E+00
          3.0E+00         2.5E+01         2.0E+00         -1.0E+00
          1.0E+01        -1.0E+00        -1.1E+01          0.0E+00
          3.0E+00        -1.0E+00         8.0E+00          1.5E+01
          0.0E+00         0.0E+00         0.0E+00          0.0E+00
```

```
C       TO SOLVE AX = B GIVEN THE PARAMETER W AND AN ITITIAL APPROXIMATION
C       X(O):
C
C       INPUT:    THE NUMBER OF EQUATIONS AND UNKNOWNS N; THE ENTRIES
C                 A(I,J), 1<=I, J<=N, OF THE MATRIX A; THE ENTRIES B(I),
C                 1<=I<=N, OF THE INHOMOGENEOUS TERM B; THE ENTRIES XO(I),
C                 1<=I<=N, OF X(0); THE PARAMETER W; TOLERANCE TOL;
C                 MAXIMUM NUMBER OF ITERATIONS N.
C
C       OUTPUT:   THE APPROXIMATE SOLUTION X(I),...,X(N) OR A MESSAGE
C                 THAT THE NUMBER OF ITERATIONS WAS EXCEEDED.
C
C       INITIALIZATION
        DIMENSION A(3,4),X1(3)
C       INPUT:  DATA FOLLOWS THE END STATEMENT
C       USE NN FOR CAPITAL N
        READ(5,1) N,TOL,NN
C       USE W FOR OMEGA
        READ(5,9) W
        M=N+1
C       B(I) = A(I,N+1) FOR 1<=I<=N
        READ(5,2) ((A(I,J),J=1,M),I=1,N)
C       USE X1 FOR XO
        READ(5,2) ( X1(I) ,I=1,N)
        WRITE(6,3)
        WRITE(6,4) ((A(I,J),J=1,M),I=1,N)
        WRITE(6,5)
        WRITE(6,4) ( X1(I) ,I=1,N)
        WRITE(6,11) W
C       STEP 1
        K=1
C       STEP 2
        WHILE(K.LE.NN)DO
            ERR=0.0
C           ERR WILL BE USED TO TEST ACCURACY, IT MEASURES THE L2 NORM
C           STEP 3
C           THE DO-LOOP COMPUTES THE SUMMATION
            DO 10 I=1,N
                S=0.0
                DO 20 J=1,N
20                  S=S-A(I,J)*X1(J)
                S=W*(S+A(I,N+1))/A(I,I)
```

```
                    ERR=ERR+S*S
C                   X IS NOT USED, SINCE ONLY ONE VECTOR IS NEEDED
10                  X1(I)=X1(I)+S
            ERR=SQRT(ERR)
            WRITE(6,6) K,ERR,(X1(I),I=1,N)
C           STEP 4
            IF(ERR.LE.TOL) THEN DO
                  WRITE(6,7) K
                  WRITE(6,12)
                  STOP
            END IF
C           STEP 5
            K=K+1
C           STEP 6 IS NOT NEEDED
      END WHILE
C     STEP 7
C     PROCEDURE COMPLETED SUCCESSFULLY
      WRITE(6,8) NN
      WRITE(6,12)
      STOP
1     FORMAT(I2,E15.8,I3)
2     FORMAT(4E15.8)
3     FORMAT('1','ORIGINAL SYSTEM: ')
4     FORMAT(('0',4(3X,E15.8)))
5     FORMAT('0','INITIAL APPROXIMATION: ')
6     FORMAT('0','ITERATION NUMBER ',I3,' GIVES ERROR ',E15.8,' FOR APPRO
     *X. ',3(2X,E15.8))
7     FORMAT('0','CONVERGENCE ON INTERATION NUMBER ',I4)
8     FORMAT('0','FAILURE AFTER INTERATION NUMBER '.I4)
9     FORMAT(E15.8)
11    FORMAT('0','OMEGA IS ',E15.8)
12    FORMAT('1')
      END
$ENTRY
 3 1.00000000E-03 30
        1.25E+00
        4.0E+00        3.0E+00        0.0E+00        24.0E+00
        3.0E+00        4.0E+00       -1.0E+00        30.0E+00
        0.0E+00       -1.0E+00        4.0E+00       -24.0E+00
        1.0E+00        1.0E+00        1.0E+00
/*
```

```
C       TO APPROXIMATE THE SOLUTION TO THE LINEAR SYSTEM AX=B WHEN A IS
C       SUSPECTED TO BE ILL-CONDITIONED:
C
C       INPUT:     THE NUMBER OF EQUATIONS AND UNKNOWNS N; THE ENTRIES
C                  A(I,J), 1<=J, J<=N, OF THE MATRIX A; THE ENTRIES B(I),
C                  1<=I<=N, OF THE INHOMOGENEOUS TERM B; THE MAXIMUM NUMBER
C                  OF ITERATIONS N; TOLERANCE TOL.
C
C       OUTPUT:    THE APPROXIMATION XX(1),...,XX(N) OR A MESSAGE THAT THE
C                  NUMBER OF ITERATIONS WAS EXCEEDED.
C
C       INITIALIZATION
        REAL*8 R(3),B(3,4),XX(3),CC
        DIMENSION A(3,4),X(3),NROW(3)
C       INPUT:   DATA FOLLOWS THE END STATEMENT
        READ(5,1) N
C       NN IS A BOUND FOR THE NUMBER OF ITERATIONS
        READ(5,1) NN
        M=N+1
        READ(5,2) ((A(I,J),J=1,M),I=1,N)
        WRITE(6,3)
        WRITE(6,4)
        WRITE(6,5) ((A(I,J),J=1,M),I=1,N)
        DO 10 I=1,N
            NROW(I)=I
            DO 10 J=1,M
10          B(I,J)=A(I,J)
C       NROW AND B HAVE BEEN INITIALIZED
C       GAUSS ELIMINATION WILL BEGIN
C       STEP 0
        KKK=N-1
        DO 21 I=1,KKK
            KK=I
            WHILE(ABS(A(KK,I)).LT.1.0E-20 .AND. KK.LE.N)DO
                KK=KK+1
            END WHILE
            IF(KK.GT.N) THEN DO
                WRITE(6,6)
                WRITE(6,3)
                STOP
            END IF
            IF(KK.NE.I) THEN DO
```

```
C                    ROW INTERCHANGE NECESSARY
                     IS=NROW(I)
                     NROW(I)=NROW(KK)
                     NROW(KK)=IS
                     DO 40 J=1,M
                          C=A(I,J)
                          A(I,J)=A(KK,J)
40                        A(KK,J)=C
            END IF
            KK=I+1
            DO 20 J=KK,N
            A(J,I)=A(J,I)/A(I,I)
            DO 20 L=KK,M
20       A(J,L)=A(J,L)-A(J,I)*A(I,L)
21       WRITE(6,5) ((A(L1,L2),L2=1,M),L1=1,N)
         IF(ABS(A(N,N)).LT.1.0E-20) THEN DO
              WRITE(6,6)
              WRITE(6,3)
              STOP
         END IF
         X(N)=A(N,M)/A(N,N)
         DO 50 I=1,KKK
              J=N-I
              JJ=J+1
              S=0
              DO 60 L=JJ,N
60            S=S-A(J,L)*X(L)
50       X(J) =(A(J,M)+S)/A(J,J)
         WRITE(6,7)
         WRITE(6,5) ((A(I,J),J=1,M),I=1,N)
         WRITE(6,8)
         WRITE(6,5) (X(I),I=1,N)
C
C     REFINEMENT BEGINS
      WRITE(6,14) (NROW(I),I=1,N)
14    FORMAT(1X,'ROW ORDER',3I2)
C
C     STEP 1
      K=1
C     INITIALIZE XX AT X
      DO 70 I=1,N
70         XX(I)=X(I)
C     STEP 2
      WHILE(K.LE.NN) DO
C          LL IS SET TO 1 IF THE DESIRED ACCURACY IN ANY COMPONENT IS
C          NOT ACHIEVED.  THUS, LL IS INITIALLY 0 FOR EACH ITERATION
           LL=0
C          STEP 3
           DO 80 I=1,N
```

```
                       R(I)=0.0D+00
C                      DO-LOOP IS TO CALCULATE SUMMATION
                       DO 90 J=1,N
90                        R(I)=R(I)-B(I,J)*XX(J)
80                     R(I)=B(I,M)+R(I)
               WRITE(6,9) K,(R(I),I=1,N)
C              STEP 4
C              SOLVES THE LINEAR SYSTEM IN THE SAME ORDER AS IN STEP 0
C              THE SOLUTION WILL BE PLACED IN X INSTEAD OF Y
               DO 100 I=1,KKK
                       I1=NROW(I)
                       KK=I+1
                       DO 100 J=KK,N
                       J1=NROW(J)
100                    R(J1)=R(J1)-A(J,I)*R(I1)
               X(N)=R(NROW(N))/A(N,N)
               DO 110 I=1,KKK
                       J=N-I
                       JJ=J+1
                       S=0
                       DO 120 L=JJ,N
120                        S=S-A(J,L)*X(L)
110                    X(J)=(R(NROW(J))+S)/A(J,J)
C              STEPS 5 AND 6
               DO 130 I=1,N
C                      IF NOT ACCURATE THEN LL=1
                       IF(ABS(X(I)).GT.1.0E-7) LL=1
130                    XX(I)=XX(I)+X(I)
               WRITE(6,11) K,(XX(I),I=1,N)
               IF(LL.EQ.0) THEN DO
                       WRITE(6,12) K
                       WRITE(6,3)
                       STOP
               END IF
C              STEP 7
               K = K+1
C                STEP 8 IS NOT USED IN THIS IMPLEMENTATION
       END WHILE
C       STEP 9
C       PROCEDURE COMPLETED UNSUCCESSFULLY
       WRITE(6,13) NN
       WRITE(6,3)
       STOP
1      FORMAT(I2)
2      FORMAT(4F10.5)
3      FORMAT('1')
4      FORMAT(1X,'ORIGINAL SYSTEM')
5      FORMAT(1X,4(5X,E15.8))
6      FORMAT(1X,'MATRIX SINGULAR - NO UNIQUE SOLUTION')
```

```
7       FORMAT( 1X, 'REDUCED SYSTEM')
8       FORMAT( 1X, 'GAUSS ELIMINATION GIVES SOLUTION-')
9       FORMAT( 1X, 'RESIDUAL NO. ', I3, ' IS', 4( 5X, D15.8))
11      FORMAT( 1X, 'ITERATE NO. ', I3, ' IS', 4( 5X, D15.8))
12      FORMAT( 1X, 'SUCCESSFUL APPROXIMATION AFTER ITERATION NO. ', I3)
13      FORMAT( 1X, 'FAILURE AFTER ', I3, ' ITERATIONS')
        END
$ENTRY
 3
25
        3.333    15920.0    -10.333    15913.0
        2.222    16.710     9.6120     28.544
       1.5611    5.1791     1.6852     8.4254
/*
```

```
C       TO COMPUTE THE COEFFICEINTS IN THE DISCRETE APPROXIMATION
C       F(X) FOR THE DATA (X(J),Y(J)), 0<=J<=2M-1 WHERE M=2**P AND
C       X(J) = -PI + J*PI/M  FOR 0 <= J <= 2M-1.
C
C       INPUT:   M; Y(0),Y(1),...,Y(2M-1).
C
C       OUTPUT:  COMPLEX NUMBERS C(0),...,C(2M-1); REAL NUMBERS
C                A(0), ..., A(M); B(1), ..., B(M-1)
C       NOTE:    THE MULTIPLICATION BY EXP(-K*PI*I) IS DONE WITHIN THE
C                THE PROGRAM.
C
        COMPLEX C(16), W(16), WW, T1, T3
        REAL Y(16)
C       INPUT:  DATA FOLLOWS THE END STATEMENT
        READ(5,1) M
        N = 2*M
C       THE Y-VALUES COULD BE INPUT HERE OR COMPUTED LATER-BUT NOT BOTH
        PI = 3.1415926
C       STEP 1
C       USE N2 FOR M, NG FOR P, NU1 FOR Q, WW FOR ZETA
        N2 = N/2
        NG = ALOG(FLOAT(N))/ALOG(FLOAT(2)) + .5
        NU1 = NG - 1
```

```
          WW = CEXP ( CMPLX( 0.0 , 2*PI/N))
C       STEP 2
C       SUBSCRIPTS ARE SHIFTED TO AVOID ZERO SUBSCRIPTS
        DO 50 I=1,N
            Z = -PI + (I-1)*PI/N2
C       THE Y-VALUES ARE COMPUTED INSTEAD OF INPUT
            Y(I) = EXP(-1-Z/PI)
50          C(I) = CMPLX( Y(I), 0.0)
C       STEP 3
        DO 40 I=1,N2
            W(I)=WW**I
40          W(N2+I)=-W(I)
C       STEP 4
        K=0
C       STEP 5
        DO 20 L=1,NG
C            STEP 6
            WHILE(K .LT. N-1) DO
C                STEP 7
                DO 30 I=1,N2
C                    STEP 8
                    M1=K/2**NU1
C                    THE SUBPROGRAM IBR DOES THE BIT REVERSAL
                    NP=IBR(M1,NG)
C                    T1 IS THE SAME AS ETA
                    T1=C(K+N2+1)
C                    STEP 9
                    IF(NP.NE.0) T1=T1*W(NP)
                    C(K+N2+1)=C(K+1)-T1
                    C(K+1)=C(K+1)+T1
C                    STEP 10
30                  K=K+1
C                STEP 11
                K=K+N2
            END WHILE
C            STEP 12
            K=0
            N2=N2/2
20          NU1=NU1-1
C       STEP 13
        WHILE(K .LT. N-1) DO
C            STEP 14
            I=IBR(K,NG)
C            STEP 15
            IF(I.GT.K) THEN DO
                T3=C(K+1)
                C(K+1)=C(I+1)
                C(I+1)=T3
            END IF
```

```
C           STEP 16
            K=K+1
      END WHILE
C     STEP 17 AND 18
      DO 60 I=1,N
            C(I)=CEXP(CMPLX(0.0,-(I-1)*PI))*C(I)/(FLOAT(N)/2)
60          WRITE(6,2) I,C(I)
      STOP
2     FORMAT(1X,I2,2(2X,E15.8))
1     FORMAT(I2)
      END
C
            FUNCTION IBR(J,NU)
            J1=J
            IBR=0
            DO 70 I=1,NU
                  J2=J1/2
                  IBR=IBR*2+(J1-2*J2)
70          J1=J2
            RETURN
            END
$ENTRY
 8
/*
```

POWER METHOD ALGORITHM   9.1

```
C     TO APPROXIMATE THE DOMINANT EIGENVALUE AND AN ASSOCIATED
C     EIGENVECTOR OF THE N BY N MATRIX A GIVEN A NONZERO VECTOR X:
C
C     INPUT:     DIMENSION N; MATRIX A; VECTOR X; TOLERANCE TOL; MAXIMUM
C                NUMBER OF ITERATIONS N.
C
C     OUTPUT:    APPROXIMATE EIGENVALUE MU; APPROXIMATE EIGENVECTOR X
C                OR A MESSAGE THAT THE MAXIMUM NUMBER OF ITERATIONS WAS
C                EXCEEDED.
C
C     INITIALIZATION
      DIMENSION A(3,3),X(3),Y(3)
C     DATA FOLLOWS THE END STATEMENT
```

```
        READ(5,1) N
        READ(5,2) ((A(I,J),J=1,N),I=1,N)
        READ(5,2) (X(I),I=1,N)
C       USE NN INSTEAD OF N FOR THE MAXIMUM NUMBER OF ITERATIONS
        READ(5,1) NN,TOL
        WRITE(6,3)
        WRITE(6,4) ((A(I,J),J=1,N),I=1,N)
        WRITE(6,5)
        WRITE(6,4) (X(I),I=1,N)
C       STEP 1
        K = 1
C       STEP 2
        LP = 1
        AMAX = ABS(X(1))
        DO 10 I=2,N
              IF(ABS(X(I)).GT.AMAX) THEN DO
                    AMAX = ABS(X(I))
                    LP = I
              END IF
10      CONTINUE
C       STEP 3
        DO 20 I=1,N
20            X(I)=X(I)/AMAX
C       STEP 4
        WHILE(K.LE.NN) DO
C             STEP 5
              DO 30 I=1,N
                    Y(I) = 0.0
                    DO 30 J=1,N
30            Y(I) = Y(I)+A(I,J)*X(J)
C             STEP 6
              YMU = Y(LP)
C             STEP 7
              LP = 1
              AMAX = ABS(Y(1))
              DO 40 I=2,N
                    IF(ABS(Y(I)).GT.AMAX) THEN DO
                          AMAX = ABS(Y(I))
                          LP = I
                    END IF
40            CONTINUE
C             STEP 8
              IF (AMAX.LT.TOL) THEN DO
                    WRITE(6,6)
                    WRITE(6,4) (X(I),I=1,N)
                    STOP
              END IF
C             STEP 9
              ERR=0.0
```

```
              DO 50 I=1,N
                  T=Y(I)/Y(LP)
                  IF(ABS(X(I)-T).GT.ERR) ERR=ABS(X(I)-T)
50                X(I) = T
              WRITE(6,7) K,YMU
              WRITE(6,8) (Y(I),I=1,N)
              WRITE(6,9) (X(I),I=1,N)
C             STEP 10
              IF(ERR.LT.TOL) THEN DO
C                 PROCEDURE COMPLETED SUCCESSFULLY
                  WRITE(6,12) YMU
                  STOP
              END IF
C             STEP 11
              K=K+1
      END WHILE
C     STEP 12
      WRITE(6,11) NN
      STOP
1     FORMAT(I2,E15.8)
2     FORMAT(3F10.5)
3     FORMAT(1X,'MATRIX A')
4     FORMAT(1X,4(5X,E15.8))
5     FORMAT(1X,'INITIAL VECTOR')
6     FORMAT(1X,'ZERO IS EIGENVALUE, PICK ANOTHER INITIAL VECTOR AND BEG
     *IN AGAIN')
7     FORMAT(1X,'ITERATION #',I3,' GIVES APPROX. = ',E15.8)
8     FORMAT(1X,'Y VECTOR IS',4(5X,E15.8))
9     FORMAT(1X,'UNIT X VECTOR IS',4(5X,E15.8))
11    FORMAT(1X,'FAILURE AFTER ITERATION #',I3)
12    FORMAT('0 EIGENVALUE IS ',E15.8)
      END
$ENTRY
 3
-4.0        14.0        0.0
-5.0        13.0        0.0
-1.0        0.0         2.0
1.0         1.0         1.0
30 1.00000000E-04
/*
```

```
C      TO APPROXIMATE THE DOMINANT EIGENVALUE AND AN ASSOCIATED
C      EIGENVECTOR OF THE N BY N SYMMETRIC MATRIX A GIVEN A NONZERO:
C      VECTOR X:
C
C      INPUT:    DIMENSION N; MATRIX A; VECTOR X; TOLERANCE TOL; MAXIMUM
C                NUMBER OF ITERATIONS N.
C
C      OUTPUT:   APPROXIMATE EIGENVALUE MU; APPROXIMATE EIGENVECTOR X OR
C                A MESSAGE THAT THE MAXIMUM NUMBER OF ITERATIONS WAS
C                EXCEEDED.
C
C      INITIALIZATION
       DIMENSION A(3,3),X(3),Y(3)
C      INPUT:   DATA FOLLOWS THE END STATEMENT
       READ(5,1) N
       READ(5,2) ((A(I,J),J=1,N),I=1,N)
C      BECAUSE OF THE WAY IN WHICH THE SUBROUTINE NORM COMPUTES THE L2
C      NORM OF A VECTOR, THE INITIAL INPUT OF X IS INTO Y
C      AND X IS INITIALIZED AT THE ZERO VECTOR
       READ(5,2) (Y(I),I=1,N)
       DO 30 I=1,N
30     X(I) = 0.0
C      USE NN IN PLACE OF N FOR THE MAXIMUM NUMBER OF ITERATIONS
       READ(5,1) NN,TOL
       WRITE(6,3)
       WRITE(6,4) ((A(I,J),J=1,N),I=1,N)
       WRITE(6,5)
       WRITE(6,4) (Y(I),I=1,N)
C      STEP 1
       K = 1
C      NORM COMPUTES THE NORM OF THE VECTOR X AND RETURNS X DIVIDED BY
C      ITS NORM IN THE VARIABLE Y
       CALL NORM(N,Y,X,ERR)
C      STEP 2
       WHILE(K.LE.NN) DO
C      STEPS 3 AND 4
            YMU=0
            DO 10 I=1,N
                 Y(I) = 0.0
                 DO 20 J=1,N
20                    Y(I) = Y(I)+A(I,J)*X(J)
10               YMU = YMU+X(I)*Y(I)
```

219

```
C               STEP 5 THIS STEP IS ACCOMPLISHED IN SUBROUTINE NORM
C               STEP 6
                CALL NORM(N, Y, X, ERR)
                WRITE(6, 7) K, YMU
                WRITE(6, 8) (Y(I), I=1, N)
                WRITE(6, 9) (X(I), I=1, N)
C               STEP 7
                IF(ERR. LT. TOL) THEN DO
                     WRITE(6, 6) YMU
C                    PROCEDURE COMPLETED SUCCESSFULLY
                     STOP
                END IF
C               STEP 8
                K = K+1
         END WHILE
C        STEP 9
         WRITE(6, 11) NN
         STOP
1        FORMAT( I2, E15. 8)
2        FORMAT( 3F10. 5)
3        FORMAT( 1X, 'MATRIX A')
4        FORMAT( 1X, 3( 5X, E15. 8))
5        FORMAT( 1X, 'INITIAL VECTOR')
6        FORMAT('0 EIGENVALUE IS ', E15. 8)
7        FORMAT( 1X, 'ITERATION NUM. ', I3, ' GIVES APPROX. = ', E15. 8)
8        FORMAT( 1X, 'Y VECTOR IS', 4( 5X, E15. 8))
9        FORMAT( 1X, 'UNIT X VECTOR IS', 4( 5X, E15. 8))
11       FORMAT( 1X, 'FAILURE AFTER ITERATION NUM. ', I3)
         END
C
                SUBROUTINE NORM(N, Y, X, ERR)
                DIMENSION X(N), Y(N)
                XL = 0. 0
                DO 10 I=1, N
10              XL = XL+Y( I)*Y( I)
                XL = SQRT(XL)
                ERR=0. 0
                IF(XL. GE. 1. 0E-20) THEN DO
                     DO 20 I=1, N
                        T=Y( I)/XL
                        IF(ABS(X( I)-T). GT. ERR) ERR=ABS(X( I)-T)
20                   X( I) = T
                     RETURN
                END IF
                WRITE(6, 1)
1               FORMAT( 1X, 'ERROR-Y IS ZERO, SO A HAS ZERO EIGENVALUE')
                STOP
                END
$ENTRY
```

220

```
 3
4.0        -1.0        1.0
-1.0        3.0        -2.0
1.0        -2.0        3.0
1.0         0.0        0.0
25 1.00000000E-04
/*
```

INVERSE POWER METHOD ALGORITHM   9.3

```
C     TO APPROXIMATE AN EIGENVALUE AND AN ASSOCIATED EIGENVECTOR OF THE
C     N BY N MATRIX A GIVEN A NONZERO VECTOR X:
C
C     INPUT:    DIMENSION N; MATRIX A; VECTOR X; TOLERANCE TOL;
C               MAXIMUM NUMBER OF ITERATIONS N.
C
C     OUTPUT:   APPROXIMATE EIGENVALUE MU; APPROXIMATE EIGENVECTOR
C               X OR A MESSAGE THAT THE MAXIMUM NUMBER OF ITERATIONS
C               WAS EXCEEDED.
C
      DIMENSION A(3,3),X(3),Y(3),NROW(3),B(3)
C     INPUT:   DATA FOLLOWS THE END STATEMENT
      READ(5,1) N
      READ(5,2) ((A(I,J),J=1,N),I=1,N)
      READ(5,2) (X(I),I=1,N)
C     USE NN INSTEAD OF N TO INDICATE THE MAXIMUM NUMBER OF ITERATIONS
      READ(5,1) NN, TOL
C     STEP 1
C     Q COULD BE INPUT INSTEAD OF COMPUTED BY DELETING THE NEXT 7 STEPS
      Q=0
      S = 0
      DO 50 I=1,N
          S= S + X(I)*X(I)
          DO 50 J=1,N
50        Q = Q + A(I,J)*X(I)*X(J)
      Q = Q/S
C     STEP 2
      K =1
      WRITE(6,4)
      WRITE(6,5) ((A(I,J),J=1,N),I=1,N)
```

221

```
      WRITE(6,6) K,(X(I),I=1,N)
      WRITE(6,7) Q
C     FORM MATRIX A - Q*I
      DO 10 I=1,N
10        A(I,I) = A(I,I)-Q
C     CALL SUBROUTINE TO COMPUTE MULTIPLIERS M(I,J) AND UPPER TRIANGULAR
C     MATRIX FOR MATRIX A USING GAUSS ELIMINATION WITH MAXIMAL COLUMN
C     PIVOTING—NROW HOLDS THE ORDERING OF ROWS FOR INTERCHANGES
      CALL MULTIP(N,A,NROW)
      WRITE(6,5) ((A(I,J),J=1,N),I=1,N)
C     STEP 3
      LP = 1
      DO 20 I=2,N
20        IF(ABS(X(I)).GT.ABS(X(LP))) LP =I
C     STEP 4
      AMAX = X(LP)
      DO 30 I=1,N
30        X(I) = X(I)/AMAX
C     STEP 5
      WHILE (K.LE.NN) DO
C         STEP 6
          DO 35 I=1,N
35        B(I)=X(I)
C         SUBROUTINE SOLVE RETURNS THE SOLUTION OF (A-Q*I)Y=B IN Y
          CALL SOLVE (N,A,NROW,B,Y)
C         STEP 7
          YMU = Y(LP)
C         STEP 8 AND 9
          LP=1
          DO 40 I=2,N
40            IF(ABS(Y(I)).GT.ABS(Y(LP))) LP=I
          AMAX = Y(LP)
          ERR=0.0
          DO 60 I=1,N
              T=Y(I)/AMAX
              IF(ABS(X(I)-T).GT.ERR) ERR=ABS(X(I)-T)
60            X(I)=T
          WRITE(6,6) K,(X(I),I=1,N),YMU,ERR
C         STEP 10
          IF (ERR.LT.TOL) THEN DO
              YMU=1/YMU+Q
              WRITE(6,3) YMU
C             PROCEDURE COMPLETED SUCCESSFULLY
              STOP
          END IF
C         STEP 11
          K = K+1
      END WHILE
C     STEP 12
```

```
      WRITE (6,9)
      STOP
1     FORMAT(I2,E15.8)
2     FORMAT(3E15.8) -
3     FORMAT('0',1X,'APPROXIMATE EIGENVALUE IS ',E15.8)
4     FORMAT('1',3X,' MATRIX A ')
5     FORMAT(3(3X,E15.8))
6     FORMAT('0',1X,I2,3(4X,E15.8),3X,'APPROX',3X,E15.8/' ERROR',E15.8)
7     FORMAT('0',3X,'Q = ',E15.8)
9     FORMAT('0',1X,'NO COVERGENCE')
      END
C
      SUBROUTINE MULTIP (N,A,NROW)
C
C     SUBROUTINE MULTIP COMPUTES THE MULTIPLIERS AND ROW ORDERING FOR
C     GAUSS ELIMINATION WITH MAXIMAL COLUMN PIVOTING-THE MULTIPLIERS
C     AND UPPER TRIANGULAR FORM ARE RETURNED IN A AND THE ROW ORDERING
C     IS RETURNED IN NROW
C
      DIMENSION A(N,N),NROW(N)
      DO 1 I=1,N
1     NROW(I) = I
      M = N-1
      DO 2 I=1,M
          IMAX = I
          J = I+1
          DO 3 IP=J,N
          L1 = NROW(IMAX)
          L2 = NROW(IP)
3         IF(ABS(A(L2,I)).GT.ABS(A(L1,I))) IMAX = IP
          IF(ABS(A(NROW(IMAX),I)).LE.1.0E-30) THEN DO
              WRITE(6,100)
              STOP
          END IF
          JJ = NROW(I)
          NROW(I) = NROW(IMAX)
          NROW(IMAX) = JJ
          I1 = NROW(I)
          DO 4 JJ=J,N
              J1 = NROW(JJ)
              A(J1,I) = A(J1,I)/A(I1,I)
              DO 5 K=J,N
5             A(J1,K) = A(J1,K)-A(J1,I)*A(I1,K)
4         CONTINUE
2     CONTINUE
      RETURN
100   FORMAT(1X,'MULTIP FAILS')
      END
C
```

```
      SUBROUTINE SOLVE(N, A, NROW, X, Y)
C
C     SUBROUTINE SOLVE ACCEPTS THE VECTOR X, THE MULTIPLIERS AND UPPER
C     TRIANGULAR FORM FOR A, THE ROW ORDERING NROW, SOLVES A*Y = X, AND
C     RETURNS THE SOLUTION IN Y
C
      DIMENSION A(N, N), NROW(N), X(N), Y(N)
      M = N-1
      DO 2 I=1, M
          J=I+1
          I1 = NROW(I)
          DO 4 JJ=J, N
              J1 = NROW(JJ)
4             X(J1) = X(J1)-A(J1, I)*X(I1)
2     CONTINUE
      IF(ABS(A(NROW(N), N)). LE. 1.0E-30) THEN DO
          WRITE(6, 100)
          STOP
      END IF
      N1 = NROW(N)
      Y(N) = X(N1)/A(N1, N)
      L = N-1
      DO 15 K=1, L
          J = L-K+1
          JJ = J+1
          N2 = NROW(J)
          Y(J)=X(N2)
          DO 16 KK=JJ, N
16            Y(J) = Y(J)-A(N2, KK)*Y(KK)
15    Y(J) = Y(J)/A(N2, J)
      RETURN
100   FORMAT(1X, 'SOLVE FAILS')
      END
$ENTRY
 3
      -4.0E+00          14.0E+00          0.0E+00
      -5.0E+00          13.0E+00          0.0E+00
      -1.0E+00           0.0E+00          2.0E+00
       1.0E+00           1.0E+00          1.0E+00
25 1.00000000E-04
/*
```

224

```
C       TO APPROXIMATE THE SECOND MOST DOMINANT EIGENVALUE AND AN
C       ASSOCIATED EIGENVECTOR OF THE N BY N MATRIX A GIVEN AN
C       APPROXIMATION LAMBDA TO THE DOMINANT EIGENVALUE, AN
C       APPROXIMATION V TO A CORRESPONDING EIGENVECTOR AND A VECTOR X
C       BELONGING TO R**(N-1):
C
C       INPUT:    DIMENSION N; MATRIX A; APPROXIMATE EIGENVALUE LAMBDA;
C                 APPROXIMATE EIGENVECTOR V BELONGING TO R**N; VECTOR X
C                 BELONGING TO R**(N-1).
C
C       OUTPUT:   APPROXIMATE EIGENVALUE MU; APPROXIMATE EIGENVECTOR U OR
C                 A MESSAGE THAT THE METHOD FAILS.
C
C       INITIALIZATION
        DIMENSION A(3,3),B(2,2),X(2),V(3),W(3),Y(2),VV(3)
C       INPUT:  DATA FOLLOWS THE END STATEMENT
        READ(5,1) N
        M = N-1
        READ(5,2) ((A(I,J),J=1,N),I=1,N)
        READ(5,2) (V(I),I=1,N)
        READ(5,4) XMU
        READ(5,3) (X(I),I=1,M)
        READ(5,1) NN,TOL
        WRITE(6,5)
        WRITE(6,6) ((A(I,J),J=1,N),I=1,N)
        WRITE(6,7)
        WRITE(6,6) (V(I),I=1,N)
        WRITE(6,8) XMU
        WRITE(6,9) (X(I),I=1,M)
C       STEP 1
        I=1
        AMAX=ABS(V(1))
        DO 100 J=2,N
             IF(ABS(V(J)).GT.AMAX) THEN DO
                  I=J
                  AMAX=ABS(V(J))
             END IF
100     CONTINUE
        IF (AMAX.LT.1.0E-20) THEN DO
             WRITE(6,10)
             STOP
        END IF
```

```
          I1=I-1
          N1=N-1
C         STEP 2
          IF (I .NE. 1) THEN DO
               DO 20 K=1, I1
                    DO 20 J=1, I1
20        B(K, J) = A(K, J)-V(K)*A(I, J)/V(I)
          END IF
C         STEP 3
          IF(I.NE.1 .AND. I.NE.N) THEN DO
               DO 30 K=I, N1

               DO 30 J=1, I1
                    B(K, J) = A(K+1, J)-V(K+1)*A(I, J)/V(I)
30        B(J, K) = A(J, K+1)-V(J)*A(I, K+1)/V(I)
          END IF
C         STEP 4
          IF (I.NE.N) THEN DO
               DO 40 K=I, N1
                    DO 40 J=I, N1
40        B(K, J) = A(K+1, J+1)-V(K+1)*A(I, J+1)/V(I)
          END IF
          WRITE(6, 11)
          WRITE(6, 12) ((B(L1, L2), L2=1, M), L1=1, M)
C         STEP 5
          CALL POWER(M, B, X, Y, YMU, NN, TOL)
C         Y IS USED IN PLACE OF W'
C         STEP 6
          IF (I.NE.1) THEN DO
               DO 50 K=1, I1
50        W(K) = Y(K)
          END IF
C         STEP 7
          W(I) = 0
C         STEP 8
          IF (I.NE.N) THEN DO
               I2 = I+1
               DO 60 K=I2, N
60        W(K) = Y(K-1)
          END IF
C         STEP 9
          DO 70 K=1, N
               S = 0
               DO 80 J=1, N
80        S = S+A(I, J)*W(J)
               S = S/V(I)
C         COMPUTE EIGENVECTOR
C         VV IS USED IN PLACE OF U HERE
70        VV(K) = (YMU-XMU)*W(K)+S*V(K)
          WRITE(6, 13) YMU
```

```
          WRITE(6, 14) (Y(I), I=1, M)
          WRITE(6, 15) (VV(I), I=1, N)
          STOP
1         FORMAT( I2, E15. 8)
2         FORMAT( 3F10. 5)
3         FORMAT( 2F10. 5)
4         FORMAT( F10. 5)
5         FORMAT( 1X, 'ORIGINAL MATRIX A')
6         FORMAT( 1X, 3( 3X, F10. 5))
7         FORMAT( 1X, 'APPROXIMATE EIGENVECTOR')
8         FORMAT( 1X, 'APPROX. DOMINANT EIGENVALUE', 3X, E15. 8)
9         FORMAT( 1X, 'INITIAL APPROX. ', 2( 3X, F10. 5))
10        FORMAT( 1X, 'INPUT ERROR')
11        FORMAT( 1X, 'REDUCED MATRIX B')
12        FORMAT( 1X, 2( 3X, E15. 8))
13        FORMAT( 1X, 'APPROX. TO SECOND EIGENVALUE', 3X, E15. 8)
14        FORMAT( 1X, 'EIGENVECTOR FOR B', 2( 3X, E15. 8))
15        FORMAT( 1X, 'EIGENVECTOR FOR A', 3( 3X, E15. 8))
          END
C
          SUBROUTINE POWER(N, A, X, Y, YMU, NN, TOL)
C         SEE ALGORITHM 8.5
C         NOTATION: N X N MATRIX A, ITERATION VECTOR Y, NORMALIZED
C         APPROX. EIGENVECTOR X, APPROX. EIGENVALUE XMU
          DIMENSION A(N, N), X(N), Y(N)
          K = 1
          LP = 1
          AMAX = ABS(X(1))
          DO 10 I=2, N
                IF(ABS(X(I)). GT. AMAX) THEN DO
                      AMAX = ABS(X(I))
                      LP = I
                END IF
10        CONTINUE
          DO 20 I=1, N
20        X(I)=X(I)/AMAX
          WHILE (K. LE. NN) DO
                DO 30 I=1, N
                      Y(I) = 0. 0
                      DO 30 J=1, N
30        Y(I) = Y(I)+A(I, J)*X(J)
          YMU=Y(LP)
          LP = 1
          AMAX = ABS(Y(1))
          DO 40 I=2, N
                IF(ABS(Y(I)). GT. AMAX) THEN DO
                      AMAX = ABS(Y(I))
                      LP = I
                END IF
```

```
40          CONTINUE
            IF (AMAX.LE.TOL) THEN DO
                  WRITE(6,100)
                  STOP
            END IF
            ERR=0.0
            DO 50 I=1,N
                T=Y(I)/Y(LP)
                IF(ABS(X(I)-T).GT.ERR) ERR=ABS(X(I)-T)
50          X(I) = T
            IF(ERR.LE.TOL) THEN DO
                  DO 60 I=1,N
60                Y(I)=X(I)
                  RETURN
            END IF
            K=K+1
       END WHILE
       WRITE(6,101)
       STOP
100    FORMAT(1X,'ZERO VECTOR-ERROR')
101    FORMAT(1X,'DIVERGENCE-ERROR')
       END
$ENTRY
 3
4.0         -1.0        1.0
-1.0         3.0        -2.0
1.0         -2.0        3.0
1.0         -1.0        1.0
6.0
2.5         -3.0
25  1.00000000E-04
/*
$JOB
```

```
C      TO OBTAIN A SYMMETRIC TRIDIAGONAL MATRIX A(N-1) SIMILAR
C      TO THE SYMMETRIC MATRIX A=A(1), CONSTRUCT THE FOLLOWING
C      MATRICES A(2),A(3),...,A(N-1) WHERE A(K) = A(I,J)**K, FOR
C      EACH K = 1,2,...,N-1:
C
C      INPUT:   DIMENSION N; MATRIX A.
C
C      OUTPUT:  A(N-1) (COULD OVER-WRITE A).
C
C      INITIALIZATION
       DIMENSION A(4,4),U(4),V(4),Z(4)
C      INPUT:   DATA FOLLOWS THE END STATEMENT
       READ(5,1) N
       READ(5,2) ((A(I,J),J=1,N),I=1,N)
       WRITE(6,3)
       WRITE(6,4) ((A(I,J),J=1,N),I=1,N)
C      STEP 1
       N2 = N-2
       N1 = N-1
       DO 10 K=1,N2
            I2 = K+1
            I3 = K+2
C           STEP 2
            Q = 0
            DO 20 J=I2,N
20               Q = Q+A(J,K)**2
            Q = SQRT(Q)
C           STEP 3
            IF(ABS(A(K+1,K)).LE.1.0E-40) THEN DO
                 S=Q
            ELSE DO
                 S = Q*(A(K+1,K)/ABS(A(K+1,K)))
            END IF
C           STEP 4
            RSQ = S*S+S*A(K+1,K)
C           NOTE: RSQ = 2*R**2
C           STEP 5
            V(K) = 0
C           NOTE: V(1)=...=V(K-1)=0, BUT ARE NOT NEEDED
            V(K+1) = A(K+1,K)+S
            IF(K.LT.N-1)THEN DO
                 DO 30 J=I3,N
```

```
30                       V(J) = A(J,K)
                 END IF
C                NOTE: W=V/SQRT(2*RSQ)=V/(2*R)
C                STEP 6
                 DO 40 J=K,N
                     U(J) = 0
                     DO 50 L=I2,N
50                       U(J) = U(J)+A(J,L)*V(L)
40                   U(J) = U(J)/RSQ
C                NOTE: U=A(K)*V/RSQ=A(K)*V/(2*R**2)
C                STEP 7
                 PROD = 0
                 DO 60 J=I2,N
60                   PROD = PROD+V(J)*U(J)
C                NOTE: PROD = V ** T *U = V**T *A(K)*V/(2*R**2)
C                STEP 8
                 QUO = PROD/(2*RSQ)
                 DO 70 J=K,N
70                   Z(J) = U(J)-QUO*V(J)
C                NOTE: Z=U-V**TUV/(2RSQ)=U-V**TUV/(4R**2)=U-WW**TU
C                     =A(K)*W/R-WW**T(A(K)W/R)
C                STEP 9
                 DO 80 L=I2,N1
                     L1 = L+1
C                    STEP 10
                     DO 90 J=L1,N
                         A(L,J) = A(L,J)-V(L)*Z(J)-V(J)*Z(L)
90                       A(J,L) = A(L,J)
C                    STEP 11
80                   A(L,L) = A(L,L)-2*V(L)*Z(L)
C                STEP 12
                 A(N,N) = A(N,N)-2*V(N)*Z(N)
                 IF (K.LT.N-1) THEN DO
C                    STEP 13
                     DO 100 J=I3,N
                         A(K,J) = 0
100                      A(J,K) = 0
                 END IF
C                STEP 14
                 A(K+1,K) = A(K+1,K)-V(K+1)*Z(K)
                 A(K,K+1) = A(K+1,K)
                 KK = K+1
                 WRITE(6,5) KK
                 WRITE(6,4)((A(I,J),J=1,N),I=1,N)
C        NOTE: THE OTHER ELEMENTS OF A(K+1) ARE THE SAME AS A(K)
C        A(K+1)=A(K)-VZ**T-ZV**T=(I-2WW**T)A(K)(I-2WW**T)
10       CONTINUE
C        STEP 15
C        OUTPUT HAS ALREADY BEEN PERFORMED
```

```
C     THE PROCESS IS COMPLETE, A(N-1) IS SYMMETRIC, TRIDIAGONAL AND
C     SIMILAR TO A
      STOP
1     FORMAT(I2)
2     FORMAT(4F10.5)
3     FORMAT(1X,'ORIGINAL MATRIX A=A(1)')
4     FORMAT(1X,4(3X,E15.8))
5     FORMAT(1X,'THE MATRIX A(',I2,') EQUALS')
      END
$ENTRY
 4
4.0        1.0        -2.0        2.0
1.0        2.0        0.0         1.0
-2.0       0.0        3.0         -2.0
2.0        1.0        -2.0        -1.0
/*
```

QR ALGORITHM   9.6

```
C     TO OBTAIN THE EIGENVALUES OF A SYMMETRIC, TRIDIAGONAL N BY N
C     MATRIX:
C
C                   A(1)  B(2)
C                   B(2)  A(2)  B(3)            .
C
C                       .     .     .
C
C                          .     .     .
C                       B(N-1)  A(N)  B(N)
C                               B(N)  A(N)
C
C     INPUT:    N; A(1),...,A(N) (DIAGONAL OF A); B(2),...,B(N)
C               (OFF-DIAGONAL OF A); MAXIMUM NUMBER OF ITERATIONS M;
C               TOLERANCE TOL
C
C     OUTPUT:   EIGENVALUES OF A OR RECOMMENDED SPLITTING OF A
C               OR A MESSAGE THAT THE MAXIMUM NUMBER OF
C               ITERATIONS WERE EXCEEDED.
C
      DIMENSION A(3),B(3),C(3),D(3),Q(3),R(3),S(3),X(3),Y(3),Z(3)
C     INPUT N, A(1),...,A(N), B(2),...,B(N), MM, TOL:   DATA AFTER
```

231

```
C        THE END STATEMENT
         READ(5,1) N
         READ(5,2) (A(I),I=1,N), (B(I),I=2,N)
C        MAXIMUM NUMBER OF ITERATIONS IS MM
         READ(5,1) MM, TOL
         ZERO = 1.0E-30
C        STEP 1
C        SET THE ACCUMULATED SHIFT TO ZERO
         SHIFT = 0
         K=1
C        STEP 2
         WHILE (K.LE.MM) DO
                WRITE(6,3) K
                WRITE(6,4) (A(I),I=1,N), (B(I),I=2,N)
C               TEST FOR CONVERGENCE / DEFLATION
C               TEST FOR POSSIBLE SPLIT
C               STEP 3
                IF (ABS(B(N)).LE.TOL) THEN DO
                     A(N)=A(N)+SHIFT
                     WRITE(6,6) A(N)
                     N=N-1
                END IF
                IF(ABS(B(2)).LE.TOL) THEN DO
                     A(1)=A(1)+SHIFT
                     WRITE(6,6) A(1)
                     N=N-1
                     A(1)=A(2)
                     DO 10 I=2,N
                          A(I)=A(I+1)
10                        B(I)=B(I+1)
                END IF
                M=N-1
                IF(M.GE.2) THEN DO
                     DO 20 I=3,M
                          IF(ABS(B(I)).LE.TOL) THEN DO
                               WRITE(6,5)
                               STOP
                          END IF
20                   CONTINUE
                END IF
C               STEP 4
C               COMPUTE SHIFT
                B1=-(A(N-1)+A(N))
                C1=A(N)*A(N-1)-B(N)*B(N)
                IF (ABS(C1).LE.ZERO) THEN DO
                     WRITE(6,7)
C                    STOP
                END IF
                D1=B1*B1-4*C1
```

232

```
               IF(D1.LT.0.0) THEN DO
                     WRITE(6,8)
                     STOP
               END IF
               D1 = SQRT(D1)
               IF(B1.GT.0.0) THEN DO
                     X1=-2*C1/(B1+D1)
                     X2=-(B1+D1)/2
               ELSE DO
                     X1=(D1-B1)/2
                     X2=2*C1/(D1-B1)
               END IF
C              IF N IS 2 THEN WE HAVE COMPUTED THE 2 EIGENVALUES
               IF(N.EQ.2) THEN DO
                     X1=X1+SHIFT
                     X2=X2+SHIFT
                     WRITE(6,6) X1
                     WRITE(6,6) X2
                     STOP
               END IF
C              STEP 5
C              ACCUMULATE SHIFT
               IF(ABS(A(N)-X1).GE.ABS(A(N)-X2)) X1=X2
               SHIFT=SHIFT+X1
               WRITE(6,9) X1
C              STEP 6
C              PERFORM SHIFT
               DO 30 I=1,N
30             D(I)=A(I)-X1
C              STEP 7
C              COMPUTE R(K)
               X(1)=D(1)
               Y(1)=B(2)
               DO 40 J=2,N
                     Z(J-1)=SQRT(X(J-1)*X(J-1)+B(J)*B(J))
                     C(J)=X(J-1)/Z(J-1)
                     S(J)=B(J)/Z(J-1)
                     Q(J-1)=C(J)*Y(J-1)+S(J)*D(J)
                     X(J)=C(J)*D(J)-S(J)*Y(J-1)
                     IF(J.NE.N) THEN DO
                           R(J-1)=S(J)*B(J+1)
                           Y(J)=C(J)*B(J+1)
                     END IF
40             CONTINUE
               M1=N-2
               WRITE(6,11) (Z(J),J=1,M)
               WRITE(6,11) (Q(J),J=1,M)
               IF(N.GT.2) WRITE(6,11) (R(J),J=1,M1)
               WRITE(6,11) (X(J),J=1,N)
```

```
                  WRITE(6, 11) (Y(J), J=1, M)
                  WRITE(6, 11) (C(J), J=2, N)
                  WRITE(6, 11) (S(J), J=2, N)
C                 STEP 8
C                 COMPUTE NEW A
                  Z(N)=X(N)
                  A(1)=S(2)*Q(1)+C(2)*Z(1)
                  B(2)=S(2)*Z(2)
                  IF(N. GT. 2) THEN DO
                        DO 50 J=2, M
                              A(J)=S(J+1)*Q(J)+C(J+1)*C(J)*Z(J)
50                          B(J+1)=S(J+1)*Z(J+1)
                  END IF
                  A(N)=C(N)*X(N)
                  K=K+1
            END WHILE
            WRITE(6, 12)
C           STEP 9
C           THE PROCESS IF COMPLETE
            STOP
1           FORMAT( I2, E15. 8)
2           FORMAT( 5E15. 8)
3           FORMAT( 1X, I2)
4           FORMAT( 1X, 'A-B', 7( 2X, E15. 8))
5           FORMAT( 1X, 'TIME TO SPLIT')
6           FORMAT( 1X, 'EIGENVALUE = ', E15. 8, ' DEFLATE AND MOVE ON')
7           FORMAT( 1X, 'NEARLY SINGULAR')
8           FORMAT( 1X, 'COMPLEX ROOTS')
9           FORMAT( 1X, 'SHIFT = ', E15. 8)
11          FORMAT( 7( 2X, E15. 8))
12          FORMAT( 1X, 'FAILURE')
            END
$ENTRY
  3
    3. 0000000E+00 3. 00000000E+00 3. 00000000E+00 1. 00000000E+00
    1. 00000000E+00          30 1. 00000000E-05
```

```
C       TO APPROXIMATE THE SOLUTION OF THE NONLINEAR SYSTEM F(X)=0 GIVEN
C       AN INITIAL APPROXIMATION X:
C
C       INPUT:    NUMBER N OF EQUATIONS AND UNKNOWNS; INITIAL APPROXIMATION
C                 X=(X(1),...,X(N)); TOLERANCE TOL; MAXIMUM NUMBER OF
C                 ITERATIONS N.
C
C       OUTPUT:   APPROXIMATE SOLUTION X=(X(1),...,X(N)) OR A MESSAGE
C                 THAT THE NUMBER OF ITERATIONS WAS EXCEEDED.
C
C       INITIALIZATION
        DIMENSION AA(3,4),Y(3),X(3)
C       INPUT:   DATA FOLLOWS THE END STATEMENT
C       USE AA FOR J;   NN FOR MAXIMUM NUMBER OF ITERATIONS N
        READ(5,6) N,NN,TOL
        PI = 4*ATAN(1.)
        READ(5,1) (X(I),I=1,N)
        WRITE(6,2)
C       STEP 1
        K=1
C       STEP 2
        WHILE(K.LE.NN) DO
C            STEP3
C            COMPUTE J(X)
             AA(1,1)=3.0
             AA(1,2)=X(3)*SIN(X(2)*X(3))
             AA(1,3)=X(2)*SIN(X(2)*X(3))
             AA(2,1)=2*X(1)
             AA(2,2)=-162*(X(2)+.1)
             AA(2,3)=COS(X(3))
             AA(3,1)=-X(2)*EXP(-X(1)*X(2))
             AA(3,2)=-X(1)*EXP(-X(1)*X(2))
             AA(3,3)=20
C            COMPUTE -F(X)
             AA(1,4)=-(3*X(1)-COS(X(2)*X(3))-.5)
             AA(2,4)=-(X(1)**2-81*(X(2)+.1)**2+SIN(X(3))+1.06)
             AA(3,4)=-(EXP(-X(1)*X(2))+20*X(3)+(10*PI-3)/3)
C            STEP 4
C            SOLVES THE N X N LINEAR SYSTEM
             CALL LIN(N,N+1,AA,Y)
C            STEPS 5 AND 6
C            R = INFINITY NORM OF Y
```

```
                R=0
                DO 10 I=1,N
                     IF(ABS(Y(I)).GT.R) R=ABS(Y(I))
10                   X(I)=X(I)+Y(I)
                WRITE(6,3) K,(X(I),I=1,N),R
C               STEP 6
                IF(R.LT.TOL) THEN DO
C                    PROCESS IS COMPLETE
                     WRITE(6,4)
                     WRITE(6,2)
                     STOP
                END IF
C               STEP 7
                K=K+1
          END WHILE
C         STEP 8
C         DIVERGENCE
          WRITE(6,5) NN
          WRITE(6,2)
          STOP
1         FORMAT(3F10.5)
2         FORMAT('1')
3         FORMAT(1X,'ITER.=',1X,I2,1X,'APPROX. SOL. IS',1X,3(1X,E15.8),1X,'A
     *PPROX. ERROR IS',1X,E15.8)
4         FORMAT(1X,'SUCCESS WITHIN TOLERANCE 1.0E-05')
5         FORMAT(1X,'DIVERGENCE - STOPPED AFTER ITER.',1X,I2)
6         FORMAT(I2,I2,E15.8)
          END
C
                SUBROUTINE LIN(N,M,A,X)
                DIMENSION A(N,M),X(N)
                K=N-1
                DO 10 I=1,K
                     Y=ABS(A(I,I))
                     IR=I
                     IA=I+1
                     DO 20 J=IA,N
                          IF(ABS(A(J,I)).GT.Y) THEN DO
                               IR=J
                               Y=ABS(A(J,I))
                          END IF
20                        CONTINUE
                     IF(Y.EQ.0.0) THEN DO
                          WRITE(6,1)
                          WRITE(6,2)
                          STOP
                     END IF
                     IF(IR.NE.I) THEN DO
                          DO 30 J=I,M
```

```
                    C=A(I,J)
                    A(I,J)=A(IR,J)
30                  A(IR,J)=C
            END IF
            DO 40 J=IA,N
                C=A(J,I)/A(I,I)
                DO 40 L=I,M
                IF(ABS(C).LT.1.0E-25) C=0
40              A(J,L)=A(J,L)-C*A(I,L)
10      CONTINUE
        IF(ABS(A(N,N)).LT.1.0E-20) THEN DO
                WRITE(6,1)
                WRITE(6,2)
                STOP
            END IF
      X(N)=A(N,M)/A(N,N)
      DO 50 I=1,K
            J=N-I
            JA=J+1
            C=A(J,M)
            DO 60 L=JA,N
60          C=C-A(J,L)*X(L)
50    X(J) = C/A(J,J)
      RETURN
1     FORMAT(1X,'LINEAR SYSTEM HAS NO SOLUTION')
2     FORMAT('1')
      END
$ENTRY
 325 1.00000000E-05
1.0         1.0         1.0
/*
```

BROYDEN ALGORITHM   10.2

```
C     TO APPROXIMATE THE SOLUTION OF THE NONLINEAR SYSTEM F(X)=0
C     GIVEN AN INITIAL APPROXIMATION X.
C
C     INPUT:    NUMBER N OF EQUATIONS AND UNKNOWNS; INITIAL
C               APPROXIMATION X=(X(1),...,X(N)); TOLERANCE TOL;
```

```
C                 MAXIMUM NUMBER OF ITERATIONS N.
C
C     OUTPUT:   APPROXIMATE SOLUTION X=(X(1),...,X(N)) OR A MESSAGE
C               THAT THE NUMBER OF ITERATIONS WAS EXCEEDED.
C
      DIMENSION A(3,3),C(3,3),X(3),Y(3),S(3),Z(3),V(3)
C     INPUT:   DATA FOLLOWS THE END STATEMENT
C     USE NN INSTEAD OF N FOR THE MAXIMUM NUMBER OF ITERATIONS
      READ(5,1) N,NN,TOL
      READ(5,2) (X(I),I=1,N)
C     INITIALIZATION
C     SN WILL REPRESENT THE L2 NORM OF THE ERROR
      SN=0
      K=0
      WRITE(6,3) K,(X(I),I=1,N),SN
C     STEP 1
C     A WILL HOLD THE JACOBIAN FOR THE INITIAL APPROXIMATION
C     THE SUBPROGRAM FP COMPUTES THE ENTRIES OF THE JACOBIAN
      DO 10 I=1,N
          DO 10 J=1,N
10        A(I,J)=FP(N,I,J,X)
C     COMPUTE V = F(X(0))
C     THE SUBPROGRAM F(N,I,X) COMPUTES THE ITH COMPONENT OF F AT X
      DO 20 I=1,N
20        V(I)=F(N,I,X)
C     STEP 2
      CALL INVER(N,A,C)
C     INVER FINDS THE INVERSE OF THE N BY N MATRIX A AND RETURNS IT IN A
C     C IS A DUMMY PARAMETER USED TO RESERVE STORAGE IN THE SUBROUTINE
C     STEP 3
      K=1
C     NOTE: S=S(1)
      CALL MULT(N,A,V,S,SN)
C     MULT COMPUTES THE PRODUCT S=-A*V AND THE L2-NORM SN OF S
      DO 30 I=1,N
30        X(I)=X(I)+S(I)
C     NOTE: X=X(1)
      WRITE(6,3) K,(X(I),I=1,N),SN
C     STEP 4
      WHILE ( K.LE.NN ) DO
C         STEP 5
C         THE VECTOR W IS NOT USED SINCE THE FUNCTION F IS EVALUATED
C         COMPONENT BY COMPONENT
          DO 40 I=1,N
              VV=F(N,I,X)
              Y(I)=VV-V(I)
40            V(I)=VV
C         NOTE: V=F(X(K)) AND Y=Y(K)
C         STEP 6
```

238

```
                  CALL MULT(N, A, Y, Z, ZN)
C                 NOTE:  Z=-A(K-1)**-1 * Y(K)
C                 STEP 7
                  P=0
C                 P WILL BE S(K)**T * A(K-1)**-1 * Y(K)
                  DO 60 I=1,N
                       P=P-S(I)*Z(I)
C                      STEP 8
                       DO 60 J=1,N
60                          C(I,J)=(S(I)+Z(I))*S(J)
                  DO 100 I=1,N
100                    C(I,I)=C(I,I)+P
C                 C=S(K)**T * A(K-1)**-1 * Y(K)*I + (S(K)+A(K-1)**-1 * Y(K))*
C                      S(K)**T )
C                 STEPS 9 AND 10
                  DO 70 I=1,N
                       DO 80 J=1,N
                            ACC=0
                            DO 90 L=1,N
C                                ACC IS THE (I,J) ENTRY OF THE INVERSE OF A
90                               ACC=ACC+C(I,L)*A(L,J)/P
C                            Z WILL HOLD THE J-TH COLUMN OF THE INVERSE OF A
80                          Z(J)=ACC
                       DO 70 J=1,N
70                          A(I,J)=Z(J)
                  CALL MULT(N, A, V, S, SN)
C                 NOTE: A=A(K)**-1 AND S=-A(K)**-1 * F(X(K))
C                 STEP 11
                  DO 110 I=1,N
110                    X(I)=X(I)+S(I)
C                 NOTE: X = X(K+1)
                  WRITE(6,3) K+1,(X(I),I=1,N),SN
C                 STEP 12
                  IF (SN.LE.TOL) THEN DO
C                      PROCEDURE COMPLETED SUCCESSFULLY
                       WRITE(6,6)
                       STOP
                  END IF
C                 STEP 13
                  K=K+1
            END WHILE
C     STEP 14
            WRITE(6,4)
            STOP
1     FORMAT(I2, I2, E15.8)
2     FORMAT(3(E15.8, 1X))
3     FORMAT(1X, I2, 3(1X, E15.8, 1X), E15.8)
4     FORMAT(1X, 'MAXIMUM NUMBER OF ITERATIONS EXCEEDED')
6     FORMAT(1X, 'PROCEDURE COMPLETED SUCCESSFULLY')
```

```
                END
C

        FUNCTION FP(N, I, J, X)
        DIMENSION X(N)
        DO CASE I
            CASE
                IF(J.EQ.1)  FP=3
                IF(J.EQ.2)  FP=X(3)*SIN(X(2)*X(3))
                IF(J.EQ.3)  FP=X(2)*SIN(X(2)*X(3))
            CASE
                IF(J.EQ.1)  FP=2*X(1)
                IF(J.EQ.2)  FP=-81*(X(2)+.1)**2
                IF(J.EQ.3)  FP= COS(X(3))
            CASE
                IF(J.EQ.1)  FP=-X(2)*EXP(-X(1)*X(2))
                IF(J.EQ.2)  FP= -X(1)*EXP(-X(1)*X(2))
                IF(J.EQ.3)  FP=20
        END CASE
        RETURN
        END
C

        FUNCTION F(N, I, X)
        DIMENSION X(N)
        IF(I.EQ.1)  F=3*X(1)-COS(X(2)*X(3))-.5
        IF(I.EQ.2)  F=X(1)**2-27*(X(2)+.1)**3+SIN(X(3))+2.00
        IF(I.EQ.3)  F=EXP(-X(1)*X(2))+20*X(3)+(10*3.141593-3)/3
        RETURN
        END
C

        SUBROUTINE MULT(N, A, Y, Z, ZN)
        DIMENSION A(N,N), Y(N), Z(N)
        ZN=0
        DO 10 I=1, N
            Z(I)=0
            DO 15 J=1, N
15          Z(I)=Z(I)-A(I, J)*Y(J)
10      ZN=ZN+Z(I)*Z(I)
        ZN=SQRT(ZN)
        RETURN
        END
C

        SUBROUTINE INVER(N, A, B)
        DIMENSION A(N,N), B(N,N)
        DO 10 I=1, N
            DO 10 J=1, N
            B(I, J)=0
10      IF(J.EQ.I)  B(I, J)=1
        DO 20 I=1, N
            I1=I+1
```

```fortran
          I2=I
          IF(I.NE.N) THEN DO
                DO 30 J=I1,N
30                IF(ABS(A(J,I)).GT.ABS(A(I2,I))) I2=J
                IF(I2.NE.I) THEN DO
                      DO 40 J=1,N
                            C=A(I,J)
                            A(I,J)=A(I2,J)
                            A(I2,J)=C
                            C=B(I,J)
                            B(I,J)=B(I2,J)
                            B(I2,J)=C
40                    CONTINUE
                END IF
          END IF
          DO 50 J=1,N
                IF (J.NE.I) THEN DO
                      C=A(J,I)/A(I,I)
                      DO 60 K=1,N
                            A(J,K)=A(J,K)-C*A(I,K)
                            B(J,K)=B(J,K)-C*B(I,K)
60                    CONTINUE
                END IF
50        CONTINUE
20    CONTINUE
      DO 70 I=1,N
            C=A(I,I)
            DO 70 J=1,N
70    A(I,J)=B(I,J)/C
      RETURN
      END
$ENTRY
 325 1.00000000E-05
 1.0E+00          1.0E+00          1.0E+00
/*
```

```
C        TO APPROXIMATE A SOLUTION P TO THE MINIMIZATION PROBLEM
C                  G(P) = MIN ( G(X) : X IN N-DIM )
C        GIVEN AN INITIAL APPROXIMATION X:
C
C        INPUT NUMBER N OF VARIABLES;  INITIAL APPROXIMATION X;
C             TOLERANCE TOL; MAXIMUM NUMBER OF ITERATIONS N.
C
C        OUTPUT APPROXIMATE SOLUTION X OR A MESSAGE OF FAILURE.
C
         LOGICAL FLAG
         DIMENSION X(3),Z(3),C(3),G(4),A(4)
         F1(X1,X2,X3)=3*X1-COS(X2*X3)-.5
         F2(X1,X2,X3)=X1*X1-81*(X2+.1)**2+SIN(X3)+1.06
         F3(X1,X2,X3)=EXP(-X1*X2)+20*X3+(10*PI-3)/3
         F(X1,X2,X3)=F1(X1,X2,X3)**2+F2(X1,X2,X3)**2+F3(X1,X2,X3)**2
C        THE FUNCTION F USED HERE CORRESPONDS TO G USED IN THE ALGORITHM
         PI=3.141593
C        INPUT:   DATA FOLLOWS THE END STATEMENT
         READ(5,1) N,M, X(1),X(2),X(3),TOL
C        STEP 1
         K=0
         WRITE(6,2) K,X(1),X(2),X(3)
         K=1
C        STEP 2
         WHILE(K.LE.M)DO
C           STEP 3
            C(1)=F1(X(1),X(2),X(3))
            C(2)=F2(X(1),X(2),X(3))
            C(3)=F3(X(1),X(2),X(3))
C           THE VECTOR C HOLDS F(X) FROM THE EXAMPLE AND G1 HOLDS G(X)
            G1=F(X(1),X(2),X(3))
            G(1)=G1
            Q=EXP(-X(1)*X(2))
            P=SIN(X(2)*X(3))
            Z(1)=6*C(1)+4*X(1)*C(2)-2*X(2)*Q*C(3)
            Z(2)=2*X(3)*P*C(1)-324*(X(2)+.1)*C(2)-2*X(1)*Q*C(3)
            Z(3)=2*X(2)*P*C(1)+2*COS(X(3))*C(2)+40*C(3)
            Z0=SQRT(Z(1)**2+Z(2)**2+Z(3)**2)
C           STEP 4
            IF(ABS(Z0).LT.1.0E-20) THEN DO
               WRITE(6,8)
```

```
              STOP
              ENDIF
C             STEP 5
              DO 10 J=1,N
10            Z(J)=Z(J)/Z0
C             A IS USED FOR ALPHA AND G0 FOR G(3)
              A(1)=0
              X0=1
              G0=F(X(1)-X0*Z(1),X(2)-X0*Z(2),X(3)-X0*Z(3))
C             STEP 6
              FLAG=.TRUE.
              WHILE(FLAG)DO
                 IF(ABS(G0).LT.ABS(G1)) FLAG=.FALSE.
C                STEPS 7 AND 8
                 X0=X0/2
                 IF(X0.LT.1.0E-10) THEN DO
                    WRITE(6,3)
                    STOP
                 ENDIF
                 G0=F(X(1)-X0*Z(1),X(2)-X0*Z(2),X(3)-X0*Z(3))
              ENDWHILE
              A(3)=X0
              G(3)=G0
C             STEP 9
              X0=X0/2
              G(2)=F(X(1)-X0*Z(1),X(2)-X0*Z(2),X(3)-X0*Z(3))
              A(2)=X0
C             STEP 10
              H1=(G(2)-G(1))/(A(2)-A(1))
              H2=(G(3)-G(2))/(A(3)-A(2))
              H3=(H2-H1)/(A(3)-A(1))
C             STEP 11
              X0=.5*(A(1)+A(2)-H1/H3)
              G0=F(X(1)-X0*Z(1),X(2)-X0*Z(2),X(3)-X0*Z(3))
C             STEP 12
              A0=X0
              DO 20 J=1,N
                IF(ABS(G(J)).LT.ABS(G0)) THEN DO
                    A0=A(J)
                    G0=G(J)
                ENDIF
20            CONTINUE
              IF(ABS(A0).LT.1.0E-10) THEN DO
                 WRITE(6,9)
                 STOP
              ENDIF
C             STEP 13
              DO 30 J=1,N
30            X(J)=X(J)-A0*Z(J)
```

```
                 WRITE(6,2) K,X(1),X(2),X(3),G0
C        STEP 14
         IF(ABS(G0).LT.TOL.OR.ABS(G0-G1).LT.TOL) THEN DO
             WRITE(6,5)
             STOP
         ENDIF
C        STEP 15
         K=K+1
      ENDWHILE
C     STEP 16
      WRITE(6,7) M
      STOP
1     FORMAT(2I2,4E15.8)
2     FORMAT(1X,I3,4(1X,E15.8))
8     FORMAT(1X,'ZERO GRADIENT: MAY HAVE MIN')
3     FORMAT(1X,'NO LIKELY IMPROVEMENT: MAY HAVE MIN')
9     FORMAT(1X,'NO CHANGE LIKELY: PROBABLY ROUNDING DIFFICULTIES')
5     FORMAT(1X,'PROCEDURE COMPLETED SUCCESSFULLY')
7     FORMAT(1X,'MAXIMUM ITERATIONS EXCEEDED')
      END
$ENTRY
 310 0.50000000E+00  0.50000000E+00  0.50000000E+00  0.00500000E+00
/*
```

LINEAR SHOOTING ALGORITHM   11.1

```
C     TO APPROXIMATE THE SOLUTION OF THE BOUNDARY-VALUE PROBLEM
C
C         -Y'' + P(X)Y' + Q(X)Y + R(X) = 0,  A <= X <= B,
C              Y(A) = ALPHA,  Y(B) = BETA
C
C     NOTE: EQUATIONS (11.5),(11.6) ARE WRITTEN AS FIRST ORDER
C           SYSTEMS AND SOLVED.
C
C     INPUT:   ENDPOINTS A,B; BOUNDARY CONDITIONS ALPHA, BETA; NUMBER OF
C              SUBINTERVALS N.
C
C     OUTPUT:  APPROXIMATIONS W(1,I) TO Y(X(I)); W(2,I) TO Y'(X(I))
C              FOR EACH I=0,1,...N.
C
```

244

```
C     VECTORS WILL NOT BE USED FOR W1 AND W2
C     INITIALIZATION
      DIMENSION U(2,10),V(2,10)
      P(X)=-2/X
      Q(X)=2/X**2
      R(X)=SIN(ALOG(X))/X**2
C     INPUT:  DATA FOLLOWS THE END STATEMENT
      READ(5,3) A,B,ALPHA,BETA,N
C     STEP 1
      H=(B-A)/N
C     U1=U(1,0), U2=U(2,0), V1=V(1,0), V2=V(2,0)
      U1=ALPHA
      U2=0.0
      V1=0.0
      V2=1.0
C     STEP 2
C     RUNGE-KUTTA METHOD FOR SYSTEMS IS USED IN STEPS 3, 4
C     THE INDEX I IS SHIFTED BY 1
      DO 10 I=1,N
C         STEP 3
          X=A+(I-1)*H
C         STEP 4
          T=X+H/2
          XK11=H*U2
          XK12=H*(P(X)*U2+Q(X)*U1+R(X))
          XK21=H*(U2+XK12/2)
          XK22=H*(P(T)*(U2+XK12/2)+Q(T)*(U1+XK11/2)+R(T))
          XK31=H*(U2+XK22/2)
          XK32=H*(P(T)*(U2+XK22/2)+Q(T)*(U1+XK21/2)+R(T))
          XK41=H*(U2+XK32)
          XK42=H*(P(X+H)*(U2+XK32)+Q(X+H)*(U1+XK31)+R(X+H))
          U1=U1+(XK11+2*XK21+2*XK31+XK41)/6
          U2=U2+(XK12+2*XK22+2*XK32+XK42)/6
          XK11=H*V2
          XK12=H*(P(X)*V2+Q(X)*V1)
          XK21=H*(V2+XK12/2)
          XK22=H*(P(T)*(V2+XK12/2)+Q(T)*(V1+XK11/2))
          XK31=H*(V2+XK22/2)
          XK32=H*(P(T)*(V2+XK22/2)+Q(T)*(V1+XK21/2))
          XK41=H*(V2+XK32)
          XK42=H*(P(X+H)*(V2+XK32)+Q(X+H)*(V1+XK31))
          V1=V1+(XK11+2*XK21+2*XK31+XK41)/6
          V2=V2+(XK12+2*XK22+2*XK32+XK42)/6
          U(1,I)=U1
          U(2,I)=U2
          V(1,I)=V1
          V(2,I)=V2
10    CONTINUE
      WRITE(6,1)
```

245

```
C       STEP 5
        W1= ALPHA
        Z=(BETA-U(1,N))/V(1,N)
        X=A
        I=0
        WRITE(6,2) I,X,W1,Z
C       STEP 6
        DO 20 I=1,N
            X=A+I*H
            W1=U(1,I)+Z*V(1,I)
            W2=U(2,I)+Z*V(2,I)
C       OUTPUT IS X(I), W(1,I), W(2,I)
20      WRITE(6,2) I,X,W1,W2
C       STEP 7
C       PROCESS IS COMPLETE
        STOP
1       FORMAT(1X,'OUTPUT:I,X(I),W(1,I),W(2,I)',/)
2       FORMAT(1X,I2,3(3X,E15.8))
3       FORMAT(4E15.8,I2)
        END
$ENTRY
 1.00000000E+00 2.00000000E+00 1.00000000E+00 2.00000000E+0010
/*
```

NONLINEAR SHOOTING ALGORITHM   11.2

```
C       TO APPROXIMATE THE SOLUTION OF THE NONLINEAR BOUNDARY-VALUE
C       PROBLEM
C
C               Y'' = F(X,Y,Y'), A<=X<=B, Y(A)=ALPHA, Y(B)=BETA:
C
C       NOTE: EQUATIONS (11.15),(11.17) ARE WRITTEN AS FIRST ORDER
C             SYSTEMS AND SOLVED
C
C       INPUT:    ENDPOINTS A,B; BOUNDARY CONDITIONS ALPHA, BETA; NUMBER OF
C                 SUBINTERVALS N; TOLERANCE TOL; MAXIMUM NUMBER OF
C                 ITERATIONS M.
C
C       OUTPUT:   APPROXIMATIONS W(1,I) TO Y(X(I)); W(2,I) TO Y'(X(I))
C                 FOR EACH I=0,1,...,N OR A MESSAGE THAT THE MAXIMUM
```

246

```
C                     NUMBER OF ITERATIONS WAS EXCEEDED.
C
C       INITIALIZATION
        DIMENSION W1(21),W2(21)
        F(X,Y,Z)=(32+2*X**3-Y*Z)/8
C       FY(X,Y,Z) REPRESENTS PARTIAL OF F WITH RESPECT TO Y
        FY(X,Y,Z)=-Z/8
C       FYP(X,Y,Z) REPRESENTS PARTIAL OF F WITH RESPECT TO Y'
        FYP(X,Y,Z)=-Y/8
C       INPUT:   DATA FOLLOWS THE END STATEMENT
        READ(5,5) A,B,ALPHA,BETA,TOL,N,M
C       STEP 1
        H=(B-A)/N
        K=1
        TK=(BETA-ALPHA)/(B-A)
C       STEP 2
        WHILE(K.LE.M) DO
C            STEP 3
C            W1(1)=W(1,0), W2(1)=W(2,0), U1=U(1), U2=U(2)
             W1(1)=ALPHA
             W2(1)= TK
             U1=0
             U2=1
C            STEP 4
C            RUNGE-KUTTA METHOD FOR SYSTEMS IS USED IN STEPS 5, 6
             DO 10 I=1,N
C                  STEP 5
                   X=A+(I-1)*H
                   T=X+H/2
C                  STEP 6
                   XK11=H*W2(I)
                   XK12=H*F(X,W1(I),W2(I))
                   XK21=H*(W2(I)+XK12/2)
                   XK22=H*F(T,W1(I)+XK11/2,W2(I)+XK12/2)
                   XK31=H*(W2(I)+XK22/2)
                   XK32=H*F(T,W1(I)+XK21/2,W2(I)+XK22/2)
                   XK41=H*(W2(I)+XK32)
                   XK42=H*F(X+H,W1(I)+XK31,W2(I)+XK32)
                   W1(I+1)=W1(I)+(XK11+2*XK21+2*XK31+XK41)/6
                   W2(I+1)=W2(I)+(XK12+2*XK22+2*XK32+XK42)/6
                   YK11=H*U2
                   YK12=H*(FY(X,W1(I),W2(I))*U1+FYP(X,W1(I),W2(I))*U2)
                   YK21=H*(U2+YK12/2)
                   YK22=H*(FY(T,W1(I),W2(I))*(U1+YK11/2)
     *                  +FYP(T,W1(I),W2(I))*(U2+YK21/2))
                   YK31=H*(U2+YK22/2)
                   YK32=H*(FY(T,W1(I),W2(I))*(U1+YK21/2)
     *                  +FYP(T,W1(I),W2(I))*(U2+YK22/2))
                   YK41=H*(U2+YK32)
```

```
                        YK42=H*(FY(X+H,W1(I),W2(I))*(U1+YK31)+FYP(X+H,W1(I),
        *               W2(I))*(U2+YK32))
                        U1=U1+(YK11+2*YK21+2*YK31+YK41)/6
                        U2=U2+(YK21+2*YK22+2*YK32+YK42)/6
10              CONTINUE
C               TEST FOR ACCURACY
C               STEP 7
                IF(ABS(W1(N+1)-BETA).LE.TOL) THEN DO
C                   PROCESS IS COMPLETE
C                   W1(J), W2(J) ARE THE APPROXIMATIONS FOR Y AND Y'
                    WRITE(6,2) K
C                   STEP 8
                    DO 20 I=1,N
                        J=I+1
                        X=A+I*H
20                      WRITE(6,1) X,W1(J),W2(J)
C                   STEP 9
                    STOP
                END IF
C               STEP 10
C               NEWTON'S METHOD APPLIED TO IMPROVE TK
                TK=TK-(W1(N+1)-BETA)/U1
                K=K+1
        END WHILE
C       STEP 11
C       METHOD FAILED
        WRITE(6,3) M
        WRITE(6,4)
        STOP
1       FORMAT(1X,3(E15.8,3X))
2       FORMAT(1X,'ORDER OF OUTPUT - X(I),W1(I),W2(I)',1X,I3,1X,'ITERATION
    *S')
3       FORMAT(1X,'METHOD FAILED AFTER ITERATION NO.',I4)
4       FORMAT('1')
5       FORMAT(5E15.8,2I2)
        END
$ENTRY
 1.00000000E+00 3.00000000E+00 1.70000000E+01 1.43333333E+01
 1.00000000E-042025
```

```
C      TO APPROXIMATE THE SOLUTION OF THE BOUNDARY-VALUE PROBLEM
C
C          Y'' = P(X)Y' + Q(X)Y + R(X), A <= X <= B
C                Y(A) = ALPHA, Y(B) = BETA:
C
C      INPUT:   ENDPOINTS A,B; BOUNDARY CONDITIONS ALPHA, BETA;
C               INTEGER N.
C
C      OUTPUT:  APPROXIMATIONS W(I) TO Y(X(I)) FOR EACH I=0,1,...N+1.
C
C      INITIALIZATION
       DIMENSION A(9),B(9),C(9),D(9),XL(9),XU(9),Z(9),W(9)
       P(X)=-2/X
       Q(X)=2/X**2
       R(X)=SIN(ALOG(X))/X**2
C      INPUT:  DATA FOLLOWS THE END STATEMENT
C      USE AA AND BB FOR ENDPOINTS A AND B
       READ(5,4) AA,BB,ALPHA,BETA,N
C      STEP 1
       H=(BB-AA)/(N+1)
       X=AA+H
       A(1)=2+H*H*Q(X)
       B(1)=-1+H*P(X)/2
       D(1)=-H*H*R(X)+(1+H*P(X)/2)*ALPHA
       M=N-1
C      STEP 2
       DO 10 I=2,M
            X=AA+I*H
            A(I)=2+H*H*Q(X)
            B(I)=-1+H*P(X)/2
            C(I)=-1-H*P(X)/2
10     D(I)=-H*H*R(X)
C      STEP 3
       X=BB-H
       A(N)=2+H*H*Q(X)
       C(N)=-1-H*P(X)/2
       D(N)=-H*H*R(X)+(1-H*P(X)/2)*BETA
C      STEP 4
C      STEPS 4 THROUGH 10 SOLVE A TRIADIAGONAL LINEAR SYSTEM USING
C      ALGORITHM 6.7
C      USE XL, XU FOR L, U RESP.
       XL(1)=A(1)
```

249

```
        XU(1)=B(1)/A(1)
C       STEP 5
        DO 20 I=2,M
            XL(I)=A(I)-C(I)*XU(I-1)
20      XU(I)=B(I)/XL(I)
C       STEP 6
        XL(N)=A(N)-C(N)*XU(N-1)
C       STEP 7
        Z(1)=D(1)/XL(1)
C       STEP 8
        DO 30 I=2,N
30      Z(I)=(D(I)-C(I)*Z(I-1))/XL(I)
C       STEP 9
        W(N)=Z(N)
C       STEP 10
        DO 40 J=1,M
            I=N-J
40      W(I)=Z(I)-XU(I)*W(I+1)
C       STEP 11
        WRITE(6,1)
        WRITE(6,2) AA,ALPHA
        DO 50 I=1,N
            X=AA+I*H
50      WRITE(6,2) X,W(I)
        WRITE(6,2) BB,BETA
        WRITE(6,3)
C       STEP 12
C       PROCEDURE IS COMPLETE
        STOP
1       FORMAT(1X,'ORDER OF OUTPUT - X(I),W(I)')
2       FORMAT(1X,2(E15.8,3X))
3       FORMAT('1')
4       FORMAT(4E15.8,I2)
        END
$ENTRY
 1.00000000E+00 2.00000000E+00 1.00000000E+00 2.00000000E+00 9
/*
```

250

```
C       TO APPROXIMATE THE SOLUTION TO THE NONLINEAR BOUNDARY-VALUE
C       PROBLEM
C
C           Y'' = F(X,Y,Y'),  A <= X <= B,  Y(A) = ALPHA,  Y(B) = BETA:
C
C       INPUT:    ENDPOINTS A,B; BOUNDARY CONDITIONS ALPHA, BETA;
C                 INTEGER N; TOLERANCE TOL; MAXIMUM NUMBER OF ITERATIONS
C                 M.
C
C       OUTPUT:   APPROXIMATIONS W(I) TO Y(X(I)) FOR EACH I=0,1,...N+1
C                 OR A MESSAGE THAT THE MAXIMUM NUMBER OF ITERATIONS WAS
C                 EXCEEDED.
C
C       INITIALIZATION
        DIMENSION W(20),A(20),C(20),D(20),XL(20),XU(20),Z(20),V(20),B(20)
        F(X,Y,Z)=(32+2*X**3-Y*Z)/8
C       FY(X,Y,Z) REPRESENTS PARTIAL OF F WITH RESPECT TO Y
        FY(X,Y,Z)=-Z/8
C       FYP(X,Y,Z) REPRESENTS PARTIAL OF F WITH RESPECT TO Y'
        FYP(X,Y,Z)=-Y/8
C       INPUT:  DATA FOLLOWS THE END STATEMENT
C       USE AA, BB FOR ENDPOINTS A, B RESP.
        READ(5,5) AA,BB,ALPHA,BETA,TOL,N,M
C       STEP 1
        N1=N-1
        H=(BB-AA)/(N+1)
C       STEP 2
        DO 10 I=1,N
10      W(I)=ALPHA + I*H*(BETA-ALPHA)/(BB-AA)
C       STEP 3
        K=1
C       STEP 4
        WHILE(K.LE.M) DO
C            STEP 5
             X=AA+H
             T=(W(2)-ALPHA)/(2*H)
             A(1)=2+H*H*FY(X,W(1),T)
             B(1)=-1+H*FYP(X,W(1),T)/2
             D(1)=-(2*W(1)-W(2)-ALPHA+H*H*F(X,W(1),T))
C            STEP 6
             DO 20 I=2,N1
                  X=AA+I*H
```

251

```
                  T=(W(I+1)-W(I-1))/(2*H)
                  A(I)=2+H*H*FY(X,W(I),T)
                  B(I)=-1+H*FYP(X,W(I),T)/2
                  C(I)=-1-H*FYP(X,W(I),T)/2
20                D(I)=-(2*W(I)-W(I+1)-W(I-1)+H*H*F(X,W(I),T))
C                 STEP 7
                  X=BB-H
                  T=(BETA-W(N-1))/(2*H)
                  A(N)=2+H*H*FY(X,W(N),T)
                  C(N)=-1-H*FYP(X,W(N),T)/2
                  D(N)=-(2*W(N)-W(N-1)-BETA+H*H*F(X,W(N),T))
C
C                 STEP 8
C                 STEPS 8 THROUGH 14 SOLVE A TRIADIAGONAL LINEAR SYSTEM
C                 USING ALGORITHM 6.7
C                 USE XL FOR L AND XU FOR U
                  XL(1)=A(1)
                  XU(1)=B(1)/A(1)
C                 STEP 9
                  DO 30 I=2,N1
                        XL(I)=A(I)-C(I)*XU(I-1)
30                XU(I)=B(I)/XL(I)
C                 STEP 10
                  XL(N)=A(N)-C(N)*XU(N-1)
C                 STEP 11
                  Z(1)=D(1)/XL(1)
C                 STEP 12
                  DO 40 I=2,N
40                Z(I)=(D(I)-C(I)*Z(I-1))/XL(I)
C                 STEP 13
                  V(N)=Z(N)
C                 VMAX IS USED TO MEASURE THE INFINITY NORM OF V
                  VMAX=ABS(V(N))
                  W(N)=W(N)+V(N)
C                 STEP 14
                  DO 50 J=1,N1
                        I=N-J
                        V(I)=Z(I)-XU(I)*V(I+1)
                        W(I)=W(I)+V(I)
                        IF(ABS(V(I)).GT.VMAX) VMAX=ABS(V(I))
50                CONTINUE
C                 STEP 15
C                 TEST FOR ACCURACY
                  IF(VMAX.LE.TOL) THEN DO
C                       STEP 16
                        WRITE(6,2) K
                        WRITE(6,3) AA,ALPHA
                        DO 60 I=1,N
                              X=AA+I*H
60                      WRITE(6,3) X,W(I)
```

252

```
                    WRITE(6,3) BB,BETA
                    WRITE(6,4)
C                   STEP 17
C                   PROCEDURE COMPLETED SUCCESSFULLY
                    STOP
                END IF
C            STEP 18
             K=K+1
        END WHILE
C       STEP 19
        WRITE(6,1) M
        WRITE(6,4)
        STOP
1       FORMAT(1X,'FAILURE AFTER ITERATION NO. ',I4)
2       FORMAT(1X,'ORDER OF OUTPUT - X(I),W(I)',1X,I3,1X,'ITERATIONS')
3       FORMAT(1X,2(E15.8,3X))
4       FORMAT('1')
5       FORMAT(5E15.8,2I2)
        END
$ENTRY
 1.00000000E+00 3.00000000E+00 1.70000000E+01 1.43333333E+01
 1.00000000E-041925
```

```
C       TO APPROXIMATE THE SOLUTION TO THE BOUNDARY-VALUE PROBLEM
C
C           -D(P(X)Y')/DX + Q(X)Y = F(X),  0 <= X <= 1,
C                   Y(0) = Y(1) = 0
C
C       WITH A PIECEWISE LINEAR FUNCTION:
C
C       INPUT    INTEGER N; MESH POINTS X(0)=0<X(1)<...<X(N)<X(N+1)=1
C
C       OUTPUT   COEFFICIENTS C(1),...,C(N) OF THE BASIS FUNCTIONS
C
C       FUNCTIONS P,Q,F DEFINED AS P,Q,FF IN SUBPROGRAMS
        DIMENSION X(22),H(21),Q(6,21)
        DIMENSION ALPHA(20),BETA(20),B(20),A(20),Z(20),ZETA(20)
```

```
       DIMENSION C(20)
C      INPUT:  DATA FOLLOWS THE END STATEMENT
       READ(5,1) N
C      SUBSCRIPTS ARE SHIFTED BY ONE TO AVOID ZERO SUBSCRIPTS
       N1 = N+1
       N2 = N+2
       NN = N-1
       READ(5,2) (X(I),I=2,N1)
       X(1)=0.
       X(N2)=1.
       WRITE(6,4)
       WRITE(6,3) (X(I),I=1,N2)
C      STEP 1
       DO 10 I=1,N1
             H(I)=X(I+1)-X(I)
10     CONTINUE
C      STEP 2
C      PEICEWISE LINEAR BASIS PHI(I) DEFINED IN SUBPROGRAMS AS NEEDED
C      STEP 3
C      COMPUTING THE INTEGRALS FOR THE ENTRIES OF THE MATRIX A
C      SIMPSON'S COMPOSITE METHOD IS USED WITHIN THE SUBPROGRAMS
       DO 20 J=2,N
           Q(1,J-1) = SIMP(1,X(J),X(J+1))/(H(J)*H(J))
           Q(2,J-1) = SIMP(2,X(J-1),X(J))/(H(J-1)*H(J-1))
           Q(3,J-1) = SIMP(3,X(J),X(J+1))/(H(J)*H(J))
           Q(4,J-1) = SIMP(4,X(J-1),X(J))/(H(J-1)*H(J-1))
           Q(5,J-1) = SIMP(5,X(J-1),X(J))/H(J-1)
           Q(6,J-1) = SIMP(6,X(J),X(J+1))/H(J)
20     CONTINUE
       Q(2,N) = SIMP(2,X(N),X(N+1))/(H(N)*H(N))
       Q(3,N) = SIMP(3,X(N+1),X(N+2))/(H(N+1)*H(N+1))
       Q(4,N) = SIMP(4,X(N),X(N+1))/(H(N)*H(N))
       Q(4,N+1) = SIMP(4,X(N+1),X(N+2))/(H(N+1)*H(N+1))
       Q(5,N) = SIMP(5,X(N),X(N+1))/H(N)
       Q(6,N) = SIMP(6,X(N+1),X(N+2))/H(N+1)
C      STEP 4
       DO 30 J = 1,NN
          ALPHA(J) = Q(4,J)+Q(4,J+1)+Q(2,J)+Q(3,J)
          BETA(J) = Q(1,J)-Q(4,J+1)
          B(J) = Q(5,J)+Q(6,J)
30     CONTINUE
C      STEP 5
       ALPHA(N) = Q(4,N)+Q(4,N+1)+Q(2,N)+Q(3,N)
       B(N) = Q(5,N)+Q(6,N)
C      STEP 6
       A(1) = ALPHA(1)
       ZETA(1) = BETA(1)/ALPHA(1)
C      STEP 7
       DO 40 J = 2,NN
```

254

```fortran
      A(J) = ALPHA(J)-BETA(J-1)*ZETA(J-1)
      ZETA(J) = BETA(J)/A(J)
40    CONTINUE
C     STEP 8
      A(N) = ALPHA(N)-BETA(N-1)*ZETA(N-1)
C     STEP 9
      Z(1) = B(1)/A(1)
C     STEP 10
      DO 50 J=2,N
         Z(J) = (B(J)-BETA(J-1)*Z(J-1))/A(J)
50    CONTINUE
C     STEP 11
      C(N) = Z(N)
C     STEP 12
      DO 60 J=1,NN
         J1 = N-J
         C(J1) = Z(J1)-ZETA(J1)*C(J1+1)
60    CONTINUE
      WRITE(6,5)
      WRITE(6,3) (C(I),I=1,N)
C     STEP 13
C     PROCESS IS COMPLETE
      STOP
1     FORMAT(I2)
2     FORMAT (5E15.8)
3     FORMAT ('0',8(1X,E15.8))
5     FORMAT ('0','OUTPUT IS C(1), C(2), ... , C(N)')
4     FORMAT ('1','OUTPUT IS X(0), X(1), ... , X(N), X(N+1)')
      END
      FUNCTION P(X)
      P = 1.0
      RETURN
      END
      FUNCTION QQ(X)
      QQ = (3.141593)**2
      RETURN
      END
      FUNCTION FF(X)
      PI = 3.141593
      FF = 2.0*PI*PI*SIN(PI*X)
      RETURN
      END
      FUNCTION SIMP(NF,A,B)
      DIMENSION Z(5)
      H = (B-A)/4.0
      DO 10 I=1,5
         Y = A+I*H
         IF(NF.EQ.1) Z(I) = (4.0-I)*I*H*H*QQ(Y)
         IF(NF.EQ.2) Z(I) = I*I*H*H*QQ(Y)
```

255

```
            IF(NF.EQ.3) Z(I) = (H*(4.0-I))**2*QQ(Y)
            IF(NF.EQ.4) Z(I) = P(Y)
            IF(NF.EQ.5) Z(I) = I*H*FF(Y)
            IF(NF.EQ.6) Z(I) = (4.0-I)*H*FF(Y)
10      CONTINUE
        SIMP = (Z(1)+Z(5)+2*Z(3)+4*(Z(2)+Z(4)))*H/3
        RETURN
        END
$ENTRY
09
  1.00000000E-01 2.00000000E-01 3.00000000E-01 4.00000000E-01
  5.00000000E-01 6.00000000E-01 7.00000000E-01 8.00000000E-01
  9.00000000E-01
```

CUBIC SPLINE RAYLEIGH-RITZ ALGORITHM   11.6

```
C       TO APPROXIMATE THE SOLUTION TO THE BOUNDARY-VALUE PROBLEM
C
C          -D(P(X)Y')/DX + Q(X)Y = F(X), 0<=X<=1, Y(0)=Y(1)=0
C
C       WITH A SUM OF CUBIC SPLINES:
C
C       INPUT    INTEGER N
C
C       OUTPUT   COEFFICIENTS C(0),...,C(N+1) OF THE BASIS FUNCTIONS
C
C       (NOTE THAT C(I) USED HERE IS 6 TIMES AS LARGE AS THE C(I)
C       USED IN THE ALGORITHM, SINCE PHI(X) IS DEFINED TO BE 1/6
C       OF THE BASIS FUNCTION USED IN THE TEXT.)
C
C       TO CHANGE PROBLEMS DO THE FOLLOWING:
C             1. CHANGE FUNCTIONS P, Q AND F IN THE SUBPROGRAMS NAMED P,Q,F
C             2. CHANGE N
C             3. CHANGE DIMENSIONS IN MAIN PROGRAM TO VALUES GIVEN BY
C                    DIMENSION A(N+2,N+3),X(N+2),C(N+2),CO(N+2,4,4),
C                               DCO(N+2,4,3),AF(N+1),BF(N+1),CF(N+1),DF(N+1),
C                               AP(N+1),BP(N+1),CP(N+1),DP(N+1),
C                               AQ(N+1),BQ(N+1),CQ(N+1),DQ(N+1)
C             4. CHANGE DIMENSIONS IN SUBROUTINE COEF TO
C                    DIMENSION AA(N+2),BB(N+2),CC(N+2),DD(N+2),
```

```
C                        H(N+2),XA(N+2),XL(N+2),XU(N+2),XZ(N+2)
C            5. CHANGE CALL STATEMENTS FOR SUBROUTINE COEF TO REFLECT
C               THE DERIVATIVES OF P,Q,F AT X(1)=0 AND X(N+2)=1
C
C
C
C      GENERAL OUTLINE
C
C            1. NODES LABELLED X(I)=(I-1)*H,  1 <= I <= N+2, WHERE
C               H=1/(N+1) SO THAT ZERO SUBSCRIPTS ARE AVOIDED
C            2. THE FUNCTIONS PHI(I) AND PHI'(I) ARE SHIFTED SO THAT
C               PHI(1) AND PHI'(1) ARE CENTERED AT X(1), PHI(2) AND PHI'(2)
C               ARE CENTERED AT X(2), . . . , PHI(N+2) AND
C               PHI'(N+2) ARE CENTERED AT X(N+2)——FOR EXAMPLE,
C                     PHI(3) = S((X-X(3))/H)
C                            = S(X/H + 2)
C            3. THE FUNCTIONS PHI(I) ARE REPRESENTED IN TERMS OF THEIR
C               COEFFICIENTS IN THE FOLLOWING WAY:
C               (PHI(I))(X) = CO(I,K,1) + CO(I,K,2)*X + CO(I,K,3)*X**2
C                          CO(I,K,4)*X**3
C               FOR X(J) <= X <= X(J+1) WHERE
C               K=1 IF J=I-2, K=2 IF J=I-1, K=3 IF J=I, K=4 IF J=I+1
C               SINCE PHI(I) IS NONZERO ONLY BETWEEN X(I-2) AND X(I+2)
C               UNLESS I = 1, 2, N+1 OR N+2
C               (SEE SUBROUTINE PHICO)
C            4. THE DERIVATIVE OF PHI(I) DENOTED PHI'(I) IS REPRESENTED
C               AS IN 3. BY ITS COEFFICIENTS DCO(I,K,L), L = 1, 2, 3
C               (SEE SUBROUTINE DPHICO).
C            5. THE FUNCTIONS P,Q AND F ARE REPRESENTED BY THEIR CUBIC
C               SPLINE INTERPOLANTS USING CLAMPLED BOUNDARY CONDITIONS
C               (SEE ALGORITHM 3.3).   THUS, FOR X(I) <= X <= X(I+1) WE
C               USE AF(I) + BF(I)*X + CF(I)*X**2 + DF(I)*X**3 TO
C               REPRESENT F(X).   SIMILARLY, AP,BP,CP,DP ARE USED FOR P AND
C               AQ,BQ,CQ,DQ ARE USED FOR Q.   (SEE SUBROUTINE COEF).
C            6. THE INTEGRANDS IS STEPS 6 AND 9 ARE REPLACED BY PRODUCTS
C               OF CUBIC POLYNOMIAL APPROXIMATIONS ON EACH SUBINTERVAL OF
C               LENGTH H AND THE INTEGRALS OF THE RESULTING POLYNOMIALS
C               ARE COMPUTED EXACTLY.   (SEE SUBROUTINE XINT).
C
C
C
       IMPLICIT REAL*8(A-H,O-Z)
       EXTERNAL P,Q,F
       DIMENSION A(11,12),X(11),C(11),CO(11,4,4),DCO(11,4,3),AF(10)
       DIMENSION DF(10),AP(10),BP(10),CP(10),DP(10),AQ(10),BQ(10),DQ(10)
       DIMENSION BF(10),CF(10),CQ(10)
       COMMON N
C      INPUT: DATA AFTER THE END STATEMENT
       READ(5,11) N
C      STEP 1
```

```
      H=1.0/(N+1)
      N1=N+1
      N2=N+2
      N3=N+3
C     INITIALIZE MATRIX A AT ZERO, NOTE THAT A(I,N+3) = B(I)
      DO 20 I=1,N2
          DO 20 J=1,N3
20                A(I,J)=0
C
C     STEP 2
C
C     X(1)=0,...,X(I)=(I-1)*H,...,X(N+2)=1
      DO 30 I=1,N2
30        X(I)=(I-1)*H
C
C     STEPS 3 AND 4 ARE IMPLEMENTED IN WHAT FOLLOWS
C
C     INITIALIZE COEFFICIENTS CO(I,J,K) AND DCO(I,J,K)
      DO 40 I=1,N2
          DO 50 J=1,4
              DO 60 K=1,4
                  CO(I,J,K)=0
                  IF(K.NE.4) THEN DO
                      DCO(I,J,K)=0
                  END IF
60            CONTINUE
C     JJ CORRESPONDS THE COEFFICIENTS OF PHI AND PHI' TO THE PROPER
C     INTERVAL INVOLVING J
              JJ=I+J-3
              CALL PHICO(I,JJ,CO(I,J,1),CO(I,J,2),CO(I,J,3),CO(I,J,4))
              CALL DPHICO(I,JJ,DCO(I,J,1),DCO(I,J,2),DCO(I,J,3))
50        CONTINUE
40    CONTINUE
C     OUTPUT THE BASIS FUNCTIONS
      WRITE(6,1)
      WRITE(6,2)
      DO 70 I=1,N2
          WRITE(6,5) I,I
          DO 70 J=1,4
              IF(I.NE.1.OR.(J.NE.1.AND.J.NE.2)) THEN DO
                  IF(I.NE.2.OR.J.NE.2) THEN DO
                      IF(I.NE.N1.OR.J.NE.4) THEN DO
                          IF(I.NE.N2.OR.(J.NE.3.AND.J.NE.4)) THEN DO
                              JJ1=I-3+J
                              JJ2=I-2+J
                              WRITE(6,3) JJ1,JJ2,(CO(I,J,K),K=1,4)
      *                             ,(DCO(I,J,K),K=1,3)
                          END IF
                      END IF
```

```
                        END IF
                END IF
70      CONTINUE
C
C       OBTAIN COEFFICIENTS FOR F, P AND Q—NOTE THAT THE 2ND AND
C       3RD ARGUMENTS ARE THE DERIVATIVES AT 0 AND 1 RESP.
C
        TTT=2*(3.141593)**3
        CALL COEF(F, TTT, -TTT, X, AF, BF, CF, DF, N2, N1)
        CALL COEF(P, 0.0D+00, 0.0D+00, X, AP, BP, CP, DP, N2, N1)
        CALL COEF(Q, 0.0D+00, 0.0D+00, X, AQ, BQ, CQ, DQ, N2, N1)
C
C       STEPS 5, 6, 7, 8, 9 ARE IMPLEMENTED IN WHAT FOLLOWS
C
        WRITE(6, 6)
        DO 80 I=1, N2
C       INDICES OF LIMITS OF INTEGRATION FOR A(I, I) AND B(I)
            J1=MIN0(I+2, N+2)
            J0=MAX0(I-2, 1)
            J2=J1-1
C       INTEGRATE OVER EACH SUBINTERVAL WHERE PHI(I) NONZERO
            DO 90 JJ=J0, J2
C       LIMITS OF INTEGRATION FOR EACH CALL
                XU=X(JJ+1)
                XL=X(JJ)
C       COEFFICIENTS OF BASES
                K=INT(I, JJ)
                A1=DCO(I, K, 1)
                B1=DCO(I, K, 2)
                C1=DCO(I, K, 3)
                D1=0
                A2=CO(I, K, 1)
                B2=CO(I, K, 2)
                C2=CO(I, K, 3)
                D2=CO(I, K, 4)
C       CALL SUBPROGRAM FOR INTEGRATION
        A(I, I)=A(I, I)+XINT(XU, XL, AP(JJ), BP(JJ), CP(JJ), DP(JJ), A1, B1, C1, D1,
       1A1, B1, C1, D1)+XINT(XU, XL, AQ(JJ), BQ(JJ), CQ(JJ), DQ(JJ), A2, B2, C2, D2,
       1A2, B2, C2, D2)
90      A(I, N+3)=A(I, N+3)+XINT(XU, XL, AF(JJ), BF(JJ), CF(JJ), DF(JJ), A2, B2, C2,
       1D2, 1.0D+00, 0.0D+00, 0.0D+00, 0.0D+00)
C       COMPUTE A(I, J) FOR J = I+1, . . . , MIN(I+3, N+2)
        K3=I+1
        IF(K3.LE.N2) THEN DO
            K2=MIN0(I+3, N+2)
            DO 100 J=K3, K2
                J0=MAX0(J-2, 1)
                DO 110 JJ=J0, J2
                    XU=X(JJ+1)
```

259

```
                          XL=X(JJ)
                          K=INT(I,JJ)
                          A1=DCO(I,K,1)
                          B1=DCO(I,K,2)
                          C1=DCO(I,K,3)
                          D1=0
                          A2=CO(I,K,1)
                          B2=CO(I,K,2)
                          C2=CO(I,K,3)
                          D2=CO(I,K,4)
                          K=INT(J,JJ)
                          A3=DCO(J,K,1)
                          B3=DCO(J,K,2)
                          C3=DCO(J,K,3)
                          D3=0
                          A4=CO(J,K,1)
                          B4=CO(J,K,2)
                          C4=CO(J,K,3)
                          D4=CO(J,K,4)
110     A(I,J)=A(I,J)+XINT(XU,XL,AP(JJ),BP(JJ),CP(JJ),DP(JJ),A1,B1,C1,D1,
       1A3,B3,C3,D3)+XINT(XU,XL,AQ(JJ),BQ(JJ),CQ(JJ),DQ(JJ),A2,B2,C2,D2,A4
       1,B4,C4,D4)
100     A(J,I)=A(I,J)
C       OUTPUT A(I,J) FOR J=I,I+1,...,MIN(I+3,N+2) AND J=N+3
        WRITE(6,7) (I,J,A(I,J),J=I,K2),I,N3,A(I,N3)
        END IF
80      CONTINUE
C
C       STEP 10
C
        DO 120 I=1,N1
            II=I+1
            DO 130 J=II,N2
                CC=A(J,I)/A(I,I)
                DO 140 K=II,N3
140             A(J,K)=A(J,K)-CC*A(I,K)
130         A(J,I)=0
120     CONTINUE
        C(N2)=A(N2,N3)/A(N2,N2)
        DO 150 I=1,N1
            J=N1-I+1
            C(J)=A(J,N3)
            JJ=J+1
            DO 160 KK=JJ,N2
160         C(J)=C(J)-A(J,KK)*C(KK)
150     C(J)=C(J)/A(J,J)
C
C       STEP 11
C
```

```
C        OUTPUT THE COEFFICIENTS C(I)
         WRITE(6,8)
         WRITE(6,9) (I,C(I),I=1,N2)
C        OBTAIN VALUES OF THE APPROX. AT THE NODES
         DO 170 I=1,N2
              S=0
              DO 180 J=1,N2
                    JO=MAX0(J-2,1)
                    J1=MIN0(J+2,N+2)
                    SS=0
                    IF(I.LT.JO.OR.I.GE.J1) THEN DO
                         S=S+C(J)*SS
                    ELSE DO
                         K=INT(J,I)
                         SS=((CO(J,K,4)*X(I)+CO(J,K,3))*X(I)+CO(J,K,2))*
     *                        X(I)+CO(J,K,1)
                         S=S+C(J)*SS
                    END IF
180      CONTINUE
         WRITE(6,10) I,X(I),S
170      CONTINUE
C
C        STEP 12
C        PROCEDURE IS COMPLETE
         STOP
1        FORMAT(1H0,28X,'BASIS FUNCTION: A + B*X + C*X**2 + D*X**3',21X,
     1   'DER. OF BASIS FUNCTION: A + B*X + C*X**2 ')
2        FORMAT(1H0,28X,'A',14X,'B',15X,'C',15X,'D',18X,'A',15X,'B',14X,
     1   'C')
3        FORMAT(1X,'ON  (X(',I2,'),X(',I2,'))',5X,4(D13.6,2X),3X,3(D13.6,
     1   2X))
5        FORMAT(1H0,'PHI(',I2,')',77X,'PHI''(',I2,')')
6        FORMAT(1H0,50X,'NONZERO ENTRIES IN MATRIX A')
7        FORMAT(1X,5('A(',I2,',',I2,') = ',D15.8))
8        FORMAT(1H0,40X,'SOLUTION IS C(1)*PHI(1) + ... + C(N+2)*PHI(N+2)
     1   WHERE ')
9        FORMAT(5(1X,'C(',I2,') = ',D15.8))
10       FORMAT(1X,'VALUE OF APPROX. AT X(',I2,') = ',D15.8,3X,'IS ',D15.8)
11       FORMAT(I2)
         END
C
         SUBROUTINE COEF(F,FPO,FPN,X,A,B,C,D,N,M)
C
C        THIS IMPLEMENTS ALGORITHM 3.3—CLAMPED BOUNDARY CUBIC SPLINE
C        F = FUNCTION TO BE APPROXIMATED, FPO = F'(0), FPN = F'(1),
C        X IS THE SET OF NODES, A,B,C,D ARE THE COEFFICIENTS TO BE
C        RETURNED, N = NUMBER OF NODES, M = NUMBER OF SUBINTERVALS
C        THE APPROX. WILL BE
C             A(I) + B(I)*X + C(I)*X**2 + D(I)*X**3 ON (X(I),X(I+1))
```

261

```
            IMPLICIT REAL*8(A-H, O-Z)
            DIMENSION AA(11), BB(11), CC(11), DD(11)
            DIMENSION X(N), A(M), B(M), C(M), D(M), H(11), XA(11), XL(11), XU(11)
            DIMENSION XZ(11)
            DO 10 I=1, M
                AA(I)=F(X(I))
10              H(I+1)=X(I+1)-X(I)
            AA(N)=F(X(N))
            XA(1)=3*(AA(2)-AA(1))/H(2)-3*FPO
            XA(N)=3*FPN-3*(AA(N)-AA(N-1))/H(N)
            XL(1)=2*H(2)
            XU(1)=.5D+00
            XZ(1)=XA(1)/XL(1)
            DO 20 I=2, M
                XA(I)=3*(AA(I+1)*H(I)-AA(I)*(X(I+1)-X(I-1))+AA(I-1)*H(I+1))
     *          /(H(I+1)*H(I))
                XL(I)=2*(X(I+1)-X(I-1))-H(I)*XU(I-1)
                XU(I)=H(I+1)/XL(I)
20              XZ(I)=(XA(I)-H(I)*XZ(I-1))/XL(I)
            XL(N)=H(N)*(2-XU(N-1))
            XZ(N)=(XA(N)-H(N)*XZ(N-1))/XL(N)
            CC(N)=XZ(N)
            DO 30 I=1, M
                J=N-I
                CC(J)=XZ(J)-XU(J)*CC(J+1)
                BB(J)=(AA(J+1)-AA(J))/H(J+1)-H(J+1)*(CC(J+1)+2*CC(J))/3
30              DD(J)=(CC(J+1)-CC(J))/(3*H(J+1))
            DO 40 I=1, M
                A(I)=(((-DD(I)*X(I))+CC(I))*X(I)-BB(I))*X(I)+AA(I)
                B(I)=(3*DD(I)*X(I)-2*CC(I))*X(I)+BB(I)
                C(I)=CC(I)-3*DD(I)*X(I)
40              D(I)=DD(I)
            RETURN
            END
C
            FUNCTION XINT (XU, XL, A1, B1, C1, D1, A2, B2, C2, D2, A3, B3, C3, D3)
C
C           FORMS THE PRODUCT (A1+B1*X+C1*X**2+D1*X**3)*(A2+B2*X+C2*X**2+
C           D2*X**3)*(A3+B3*X+C3*X**2+D3*X**3) TO OBTAIN
C           10*C(1)*X**9 + 9*C(2)*X**8 + . . . + 2*C(9)*X + C(10)
C           WHICH IS INTEGRATED FROM XL TO XU
            IMPLICIT REAL*8(A-H, O-Z)
            DIMENSION C(10)
            AA=A1*A2
            BB=A1*B2+A2*B1
            CC=A1*C2+B1*B2+C1*A2
            DD=A1*D2+B1*C2+C1*B2+D1*A2
            EE=B1*D2+C1*C2+D1*B2
            FF=C1*D2+D1*C2
```

```
                GG=D1*D2
                C(10)=AA*A3
                C(9)=(AA*B3+BB*A3)/2
                C(8)=(AA*C3+BB*B3+CC*A3)/3
                C(7)=(AA*D3+BB*C3+CC*B3+DD*A3)/4
                C(6)=(BB*D3+CC*C3+DD*B3+EE*A3)/5
                C(5)=(CC*D3+DD*C3+EE*B3+FF*A3)/6
                C(4)=(DD*D3+EE*C3+FF*B3+GG*A3)/7
                C(3)=(EE*D3+FF*C3+GG*B3)/8
                C(2)=(FF*D3+GG*C3)/9
                C(1)=(GG*D3)/10
                XHIGH=0
                XLOW=0
                DO 10 I=1,10
                    XHIGH=(XHIGH+C(I))*XU
    10              XLOW=(XLOW+C(I))*XL
                XINT=XHIGH-XLOW
                RETURN
                END
C
                SUBROUTINE PHICO(I,J,A,B,C,D)
C
C       COMPUTES PHI(I) AS A+B*X+C*X**2+D*X**3 FOR X IN (X(J),X(J+1)) BY
C       MULTIPLYING AND SIMPLIFYING THE EQUATIONS IN STEP 2
                IMPLICIT REAL*8(A-H,O-Z)
                COMMON N
                H=1.0D+00/(N+1)
                A=0
                B=0
                C=0
                D=0
                E=I-1
                IF(J.LT.I-2.OR.J.GE.I+2) RETURN
                IF(I.EQ.1.AND.J.LT.I) RETURN
                IF(I.EQ.2.AND.J.LT.I-1) RETURN
                IF(I.EQ.N+1.AND.J.GT.N+1) RETURN
                IF(I.EQ.N+2.AND.J.GE.N+2) RETURN
                IF(J.LE.I-2) THEN DO
                    A= (((-E+6)*E-12)*E+8)/24
                    B=((E-4)*E+4)/(8*H)
                    C=(-E+2)/(8*H*H)
                    D=1/(24*H**3)
                    RETURN
                ELSE DO
                    IF(J.GT.I-1.AND.J.GT.I) THEN DO
                        A=(((E+6)*E+12)*E+8)/24
                        B=((-E-4)*E-4)/(8*H)
                        C=(E+2)/(8*H*H)
                        D=-1/(24*H**3)
```

```
                    RETURN
            ELSE DO
                IF(J.GT.I-1) THEN DO
                        A=((-3*E-6)*E*E+4)/24
                        B=(3*E+4)*E/(8*H)
                        C=(-3*E-2)/(8*H*H)
                        D=1/(8*H**3)
                        IF(I.NE.1.AND.I.NE.N+1) RETURN
                ELSE DO
                        A=((3*E-6)*E*E+4)/24
                        B=(-3*E+4)*E/(8*H)
                        C=(3*E-2)/(8*H*H)
                        D=-1/(8*H**3)
                        IF(I.NE.2.AND.I.NE.N+2) RETURN
                END IF
            END IF
        END IF
        IF(I.LE.2) THEN DO
            AA=1.0D+00/24
            BB=-1/(8*H)
            CC=1/(8*H*H)
            DD=-1/(24*H**3)
            IF(I.EQ.2) THEN DO
                    A=A-AA
                    B=B-BB
                    C=C-CC
                    D=D-DD
                    RETURN
            ELSE DO
                    A=A-4*AA
                    B=B-4*BB
                    C=C-4*CC
                    D=D-4*DD
                    RETURN
            END IF
        ELSE DO
            EE=N+2
            AA=(((-EE+6)*EE-12)*EE+8)/24
            BB=((EE-4)*EE+4)/(8*H)
            CC=(-EE+2)/(8*H*H)
            DD=1/(24*H**3)
            IF(I.EQ.N+1) THEN DO
                    A=A-AA
                    B=B-BB
                    C=C-CC
                    D=D-DD
                    RETURN
            ELSE DO
                    A=A-4*AA
```
```
                        264
```

```
                  B=B-4*BB
                  C=C-4*CC
                  D=D-4*DD
            END IF
      END IF
      RETURN
      END
C
      SUBROUTINE DPHICO(I,J,A,B,C)
C
C     SAME AS PHICO EXCEPT THE COEFFICIENTS ARE FOR PHI'(I)
      IMPLICIT REAL*8(A-H,O-Z)
      COMMON N
      H=1.0D+00/(N+1)
      A=0
      B=0
      C=0
      E=I-1
      IF(J.LT.I-2 .OR. J.GE.I+2) RETURN
      IF(I.EQ.1 .AND. J.LT.I) RETURN
      IF(I.EQ.2 .AND. J.LT.I-1) RETURN
      IF(I.EQ.N+1 .AND. J.GT.N+1) RETURN
      IF(I.EQ.N+2 .AND. J.GE.N+2) RETURN
      IF(J.LE.I-2) THEN DO
            A=((E-4)*E+4)/(8*H)
            B=(-E+2)/(4*H*H)
            C=1/(8*H**3)
            RETURN
      ELSE DO
            IF(J.GT.I-1.AND.J.GT.I) THEN DO
                  A=((-E-4)*E-4)/(8*H)
                  B=(E+2)/(4*H*H)
                  C=-1/(8*H**3)
                  RETURN
            ELSE DO
                  IF(J.GT.I-1) THEN DO
                        A=(3*E+4)*E/(8*H)
                        B=(-3*E-2)/(4*H*H)
                        C=3/(8*H**3)
                        IF(I.NE.1 .AND.I.NE.N+1) RETURN
                  ELSE DO
                        A=(-3*E+4)*E/(8*H)
                        B=(3*E-2)/(4*H*H)
                        C=-3/(8*H**3)
                        IF(I.NE.2 .AND.I.NE.N+2) RETURN
                  END IF
            END IF
      END IF
      IF(I.LE.2) THEN DO
```

```
                  AA=-1/(8*H)
                  BB=1/(4*H*H)
                  CC=-1/(8*H**3)
                  IF(I.EQ.2) THEN DO
                       A=A-AA
                       B=B-BB
                       C=C-CC
                       RETURN
                  ELSE DO
                       A=A-4*AA
                       B=B-4*BB
                       C=C-4*CC
                       RETURN
                  END IF
            ELSE DO
                  EE=N+2
                  AA=((EE-4)*EE+4)/(8*H)
                  BB=(-EE+2)/(4*H*H)
                  CC=1/(8*H**3)
                  IF(I.EQ.N+1) THEN DO
                       A=A-AA
                       B=B-BB
                       C=C-CC
                       RETURN
                  ELSE DO
                       A=A-4*AA
                       B=B-4*BB
                       C=C-4*CC
                  END IF
            END IF
      END IF
      RETURN
      END
C
      FUNCTION INT(J,JJ)
C
C     CORRESPONDS THE INTERVAL (X(JJ),X(JJ+1)) TO PROPER INDEX
C     K IN CO(J,K,L),L=1,...,4 AND IN DCO(J,K,L),L=1,2,3
      IMPLICIT REAL*8(A-H,O-Z)
      IF(JJ.LE.J-3 .OR. JJ.GE.J+2) THEN DO
            WRITE(6,2)
2           FORMAT(1X,'ERROR IN INT')
            STOP
      ELSE DO
            IF(JJ.EQ.J-2) INT=1
            IF(JJ.EQ.J-1) INT=2
            IF(JJ .EQ. J) INT=3
            IF(JJ.EQ.J+1) INT=4
      END IF
      RETURN
```

```
          END
C

          FUNCTION P(X)
          IMPLICIT REAL*8(A-H,O-Z)
          P=1.0D+00
          RETURN
          END
C

          FUNCTION Q(X)
          IMPLICIT REAL*8(A-H,O-Z)
          Q=3.141593*3.141593
          RETURN
          END
C

          FUNCTION F(X)
          IMPLICIT REAL*8(A-H,O-Z)
          F=2*(3.141593)**2*DSIN(3.141593*X)
          RETURN
          END
$ENTRY
09
/*
```

POISSON EQUATION FINITE-DIFFERENCE ALGORITHM   12.1

```
C     TO APPROXIMATE THE SOLUTION TO THE POISSON EQUATION
C         DEL(U) = F(X,Y), A <= X <= B, C <= Y <= D,
C     SUBJECT TO BOUNDARY CONDITIONS:
C                     U(X,Y)=G(X,Y),
C     IF X = A OR X = B, FOR C <= Y <= D
C     IF Y = C OR Y = D, FOR A <= X <= B
C
C     INPUT:   ENDPOINTS A, B, C, D; INTEGERS M, N; TOLERENCE TOL;
C              MAXIMUM NUMBER OF ITERATIONS M.
C
C     OUTPUT:  APPROXIMATIONS W(I,J) TO U(X(I),Y(J)) FOR EACH
C              I=1,...,N-1 AND J=1,...,M-1 OR A MESSAGE THAT THE
C              MAXIMUM NUMBER OF ITERATIONS WAS EXCEEDED.
C
C     INITIALIZATION
```

```
C          DIMENSION W(N-1,M-1),X(N-1),Y(M-1)
           DIMENSION W(5,4),X(5),Y(4)
C          USE  -F(X,Y) INSTEAD OF F(X,Y) AND WE USED EXACT SOLUTION FOR
C          BOUNDARY VALUES
           F(X,Y)=-X*EXP(Y)
           G(X,Y)= X*EXP(Y)
C          LB IS USED FOR CAPITAL M
C          INPUT:   DATA FOLLOWS THE END STATEMENT
           READ(5,4) A,B,C,D
           READ(5,5) TOL,M,N,LB
           MM=M-1
           MMM=M-2
           NN=N-1
           NNN=N-2
C          XK WILL BE USED FOR K
C          STEP 1
           H=(B-A)/N
           XK=(D-C)/M
C          A,B,C,D WILL BE USED FOR X(0),X(N),Y(0),Y(M)
C          STEPS 2, 3 CONSTRUCT MESH POINTS
C          STEP 2
           DO 10 I=1,NN
10             X(I)=A+I*H
C          STEP 3
           DO 20 J=1,MM
20             Y(J)=C+J*XK
C          STEP 4
           DO 30 I=1,NN
               DO 30 J=1,MM
30                 W(I,J)=0
C          STEP 5
C          USE V FOR LAMBDA, VV FOR MU
           V=H*H/(XK*XK)
           VV=2*(1+V)
           L=1
C          Z IS A NEW VALUE OF W(I,J) TO BE USED IN COMPUTING THE ERROR E
C          USE E INSTEAD OF NORM
C          STEP 6
           WHILE(L.LE.LB) DO
C          STEPS 7 THROUGH 20 PERFORM GAUSS-SEIDEL ITERATIONS
C              STEP 7
               Z=(H*H*F(X(1),Y(MM))+G(A,Y(MM))+V*G(X(1),D)+W(1,MM-1)*V+W(2,
      *        MM))/VV
               E=ABS(W(1,MM)-Z)
               W(1,MM)=Z
C              STEP 8
               DO 40 I=2,NNN
                   Z=(H*H*F(X(I),Y(MM))+V*G(X(I),D)+W(I-1,MM)+W(I+1,MM)+V*
      *            W(I,MM-1))/VV
                   IF(ABS(W(I,MM)-Z).GT.E) E=ABS(W(I,MM)-Z)
```
268

```
      READ(5,3) FX, FT, ALPHA, M, N
      MM=M-1
      MMM=MM-1
      NN=N-1
C     STEP 1
      H=FX/M
C     USE TK FOR K
      TK=FT/N
C     USE VV FOR LAMBDA
      VV=ALPHA*ALPHA*TK/(H*H)
C     STEP 2
      DO 10 I=1,MM
C     INITIAL VALUES
10        W(I)=F(I*H)
C     STEP 3
C     STEPS 3 THROUGH 11 SOLVE A TRIDIAGONAL LINEAR SYSTEM
C     USING ALGORITHM 6.7
C     USE XL FOR L, XU FOR U
      XL(1)=1+2*VV
      XU(1)=-VV/XL(1)
C     STEP 4
      DO 20 I=2,MMM
          XL(I)=1+2*VV+VV*XU(I-1)
20    XU(I)=-VV/XL(I)
C     STEP 5
      XL(MM)=1+2*VV+VV*XU(MMM)
C     STEP 6
      DO 30 J=1,N
C         STEP 7
C         CURRENT T(J)
          T=J*TK
          Z(1)=W(1)/XL(1)
C         STEP 8
          DO 40 I=2,MM
40        Z(I)=(W(I)+VV*Z(I-1))/XL(I)
C         STEP 9
          W(MM)=Z(MM)
C         STEP 10
          DO 50 II=1,MMM
                I=MMM-II+1
50        W(I)=Z(I) -XU(I)*W(I+1)
30    CONTINUE
C     STEP 11 IS CHANGED TO OUTPUT ONLY THE FINAL TIME VALUES
C     OUTPUT FINAL TIME VALUE
      WRITE(6,1) FT
      DO 60 I=1,MM
          X=I*H
60    WRITE(6,2) I,X,W(I)
C     STEP 12
```

```
        STOP
1       FORMAT( 1X, 'FINAL TIME VALUE IS', 1X, E15. 8, 1X, 'ORDER OF OUTPUT IS I,
       *X( I ), W( I )')
2       FORMAT( 1X, I3, 2( 3X, E15. 8))
3       FORMAT( 3E15. 8, 2I3)
        END
$ENTRY
 1. 00000000E+00  5. 00000000E-01  1. 00000000E+00  10  50
/*
```

CRANK-NICOLSON ALGORITHM · 12.3

```
C       TO APPROXIMATE THE SOLUTION TO A PARABOLIC PARTIAL-DIFFERENTIAL
C       EQUATION SUBJECT TO THE BOUNDARY CONDITIONS
C                   U( 0, T) = U( L, T) = 0,  0 < T < MAX T
C       AND THE INITIAL CONDITIONS
C                   U( X, 0) = F( X),  0 <= X <= L:
C
C       INPUT:    ENDPOINT L; MAXIMUM TIME T; CONSTANT ALPHA; INTEGERS
C                 M, N:
C
C       OUTPUT:   APPROXIMATIONS W( I, J) TO U( X( I ), T( J )) FOR EACH
C                 I=1, ..., M-1 AND J=1, ..., N.
C
C       INITIALIZATION
C       DIMENSION V( M), XL( M-1), XU( M-1), Z( M-1)
        DIMENSION V( 10), XL( 9), XU( 9), Z( 9)
        F( X)=SIN( PI*X)
        PI=3. 1415927
C       INPUT:   DATA FOLLOWS THE END STATEMENT
C       V IS USED FOR W, FT FOR CAPITAL T,  FX FOR L
        READ( 5, 3) FX, FT, ALPHA, M, N
        MM=M-1
        MMM=MM-1
        NN=N-1
C       STEP 1
        H=FX/M
C       TK IS USED FOR K
        TK=FT/N
```

272

```
C      VV IS USED FOR LAMBDA
       VV=ALPHA*ALPHA*TK/(H*H)
C      SET V(M)=0
       V(M)=0
C      STEP2
C      INITIAL VALUES
       DO 10 I=1,MM
10         V(I)=F(I*H)
C      STEP 3
C      STEPS 3 THROUGH 11 SOLVE A TRIDIAGONAL LINEAR SYSTEM
C      USING ALGORITHM 6.7
C      USE XL FOR L, XU FOR U
       XL(1)=1+VV
       XU(1)=-VV/(2*XL(1))
C      STEP 4
       DO 20 I=2,MMM
           XL(I)=1+VV+VV*XU(I-1)/2
20     XU(I)=-VV/(2*XL(I))
C      STEP 5
       XL(MM)=1+VV+VV*XU(MMM)/2
C      STEP 6
       DO 30 J=1,N
C          STEP 7
C          CURRENT T(J)
           T=J*TK
           Z(1)=((1-VV)*V(1)+VV*V(2)/2)/XL(1)
C          STEP 8
           DO 40 I=2,MM
40         Z(I)=((1-VV)*V(I)+VV/2*(V(I+1)+V(I-1)+Z(I-1)))/XL(I)
C          STEP 9
           V(MM)=Z(MM)
C          STEP 10
           DO 50 II=1,MMM
                I=MMM-II+1
50         V(I)=Z(I) -XU(I)*V(I+1)
30     CONTINUE
C      STEP 11—OUTPUT WILL BE ONLY FOR T=FT
       WRITE(6,1) FT
       DO 60 I=1,MM
           X=I*H
60     WRITE(6,2) I,X,V(I)
C      STEP 12
C      PROCEDURE IS COMPLETE
       STOP
1      FORMAT(1X,'OUTPUT FOR TIME = ' ,1X,E15.8,1X,'ORDER OF OUTPUT IS I,
      *X(I),V(I)')
2      FORMAT(1X,I3,2(3X,E15.8))
3      FORMAT(3E15.8,2I3)
       END
```

$ENTRY
  1.00000000E+00 5.00000000E-01 1.00000000E+00 10 50
/*

WAVE EQUATION FINITE-DIFFERENCE ALGORITHM   12.4

```
C       TO APPROXIMATE THE SOLUTION TO THE WAVE EQUTION:
C       SUBJECT TO THE BOUNDARY CONDITIONS
C               U(0,T) = U(L,T) = 0, 0 < T < MAX T
C       AND THE INITIAL CONDITIONS
C           U(X,0) = F(X) AND DU(X,0)/DT = G(X), 0 <= X <= L:
C
C       INPUT:    ENDPOINT L; MAXIMUM TIME T; CONSTANT ALPHA; INTEGERS M,N.
C
C       OUTPUT:   APPROXIMATIONS W(I,J) TO U(X(I),T(J)) FOR EACH I=0,...M
C                 AND J=0,..,N.
C
C       INITIALIZATION
C       DIMENSION W(M+1,N+1)
        DIMENSION W(11,21)
        F(X)=SIN(PI*X)
        G(X)=0
        PI=3.141593
C       INPUT:  DATA FOLLOWS THE END STATEMENT
C       FT IS CAPITAL T AND FX IS L
        READ(5,3) FX,FT,ALPHA,M,N
        M1=M+1
        N1=N+1
        MM=M-1
        NN=N-1
C       STEP 1
C       TK IS USED FOR K, V FOR LAMBDA
        H=FX/M
        TK=FT/N
        V=TK*ALPHA/H
C       STEP 2
C       THE SUBSCRIPTS ARE SHIFTED TO AVOID ZERO SUBSCRIPTS
        DO 20 J=2.N1
            W(1,J)=0
20      W(M1,J)=0
C       STEP 3
```

274

```
      W(1,1)=F(0.0)
      W(M1,1)=F(FX)
C     STEP 4
      DO 30 I=2,M
         W(I,1)=F((I-1)*H)
30    W(I,2)=(1-V*V)*F((I-1)*H)+V*V*(F(I*H)+F((I-2)*H))/2+TK*G((I-1)*H)
C     STEP 5
      DO 40 J=2,N
         DO 40 I=2,M
40    W(I,J+1)=2*(1-V*V)*W(I,J)+V*V*(W(I+1,J)+W(I-1,J))-W(I,J-1)
C     STEP 6
C     OUTPUT ONLY FOR TIME VALUE T=FT
      WRITE(6,1) FT
      DO 50 I=1,M1
         X=(I-1)*H
50    WRITE(6,2) I,X,W(I,N1)
C     STEP 7
C     PROCEDURE IS COMPLETE
      STOP
1     FORMAT(1X,'FINAL TIME VALUE IS',1X,E15.8,1X,'ORDER OF OUTPUT IS I,
     *X(I),W(I,N)')
2     FORMAT(1X,I3,2(3X,E15.8))
3     FORMAT(3E15.8,2I3)
      END
$ENTRY
 1.00000000E+00 1.00000000E+00 2.00000000E+00 10 20
/*
```

FINITE-ELEMENT ALGORITHM   12.5

```
C     TO APPROXIMATE THE SOLUTION TO AN ELLIPTIC PARTIAL-DIFFERENTIAL
C     EQUATION SUBJECT TO DIRICHLET, MIXED, OR NEUMANN BOUNDARY
C     CONDITIONS:
C
C     INPUT    SEE STEP 0
C
C     OUTPUT   DESCRIPTION OF TRIANGLES, NODES, LINE INTEGRALS, BASIS
C              FUNCTIONS, LINEAR SYSTEM TO BE SOLVED, AND THE
C              COEFFICIENTS OF THE BASIS FUNCTIONS
C
```

```
C
C      STEP 0
C      GENERAL OUTLINE
C
C         1. TRIANGLES NUMBERED: 1 TO K FOR TRIANGLES WITH NO EDGES ON
C            SCRIPT-S-1 OR SCRIPT-S-2, K+1 TO N FOR TRIANGLES WITH
C            EDGES ON SCRIPT-S-2, N+1 TO M FOR REMAINING TRIANGLES.
C            NOTE: K=0 IMPLIES THAT NO TRIANGLE IS INTERIOR TO D.
C            NOTE: M=N IMPLIES THAT ALL TRIANGLES HAVE EDGES ON
C            SCRIPT-S-2.
C
C         2. NODES NUMBERED: 1 TO LN FOR INTERIOR NODES AND NODES ON
C            SCRIPT-S-2, LN+1 TO LM FOR NODES ON SCRIPT-S-1.
C            NOTE: LM AND LN REPRESENT LOWER CASE M AND N RESP.
C            NOTE: LN=LM IMPLIES THAT SCRIPT-S-1 CONTAINS NO NODES.
C            NOTE: IF A NODE IS ON BOTH SCRIPT-S-1 AND SCRIPT-S-2, THEN
C            IT SHOULD BE TREATED AS IF IT WERE ONLY ON SCRIPT-S-1.
C
C         3. NL=NUMBER OF LINE SEGMENTS ON SCRIPT-S-2
C            LINE(I,J) IS AN NL BY 2 ARRAY WHERE
C            LINE(I,1)= FIRST NODE ON LINE I AND
C            LINE(I,2)= SECOND NODE ON LINE I TAKEN
C            IN POSITIVE DIRECTION ALONG LINE I
C
C         4. FOR THE NODE LABELLED KK,KK=1,...,LM WE HAVE:
C            A) COORDINATES XX(KK),YY(KK)
C            B) NUMBER OF TRIANGLES IN WHICH KK IS A VERTEX= LL(KK)
C            C) II(KK,J) LABELS THE TRIANGLES KK IS IN AND
C               NV(KK,J) LABELS WHICH VERTEX NODE KK IS FOR
C               EACH J=1,...,LL(KK)
C
C         5. NTR IS AN M BY 3 ARRAY WHERE
C            NTR(I,1)=NODE NUMBER OF VERTEX 1 IN TRIANGLE I
C            NTR(I,2)=NODE NUMBER OF VERTEX 2 IN TRIANGLE I
C            NTR(I,3)=NODE NUMBER OF VERTEX 3 IN TRIANGLE I
C
C         6. FUNCTION SUBPROGRAMS:
C            A) P,Q,R,F,G,G1,G2 ARE THE FUNCTIONS SPECIFIED BY
C               THE PARTICULAR DIFFERENTIAL EQUATION
C            B) RR IS THE INTEGRAND
C               R*N(J)*N(K) ON TRIANGLE I IN STEP 4
C            C) FFF IS THE INTEGRAND F*N(J) ON TRIANGLE I IN STEP 4
C            D) GG1=G1*N(J)*N(K)
C               GG2=G2*N(J)
C               GG3=G2*N(K)
C               GG4=G1*N(J)*N(J)
C               GG5=G1*N(K)*N(K)
C               INTEGRANDS IN STEP 5
C            E) QQ(FF) COMPUTES THE DOUBLE INTEGRAL OF ANY
```

```
C                   INTEGRAND FF OVER A TRIANGLE WITH VERTICES GIVEN BY
C                   NODES J1, J2, J3 - THE METHOD IS AN O(H**2) APPROXIMATION
C                   FOR TRIANGLES
C               F)  SQ(PP) COMPUTES THE LINE INTEGRAL OF ANY INTEGRAND PP
C                   ALONG THE LINE FROM (XX(J1), YY(J1)) TO (XX(J2), YY(J2))
C                   BY USING A PARAMETERIZATION GIVEN BY:
C                      X=XX(J1)+(XX(J2)-XX(J1))*T
C                      Y=YY(J1)+(YY(J2)-YY(J1))*T
C                   FOR 0 <= T <= 1
C                   AND APPLYING SIMPSON'S COMPOSITE METHOD WITH H=.01
C
C           7. ARRAYS:
C               A)  A, B, C ARE M BY 3 ARRAYS WHERE THE BASIS FUNCTION N
C                   FOR THE ITH TRIANGLE, JTH VERTEX IS
C                   N(X, Y)=A(I, J)+B(I, J)*X+C(I, J)*Y
C                   FOR J=1, 2, 3 AND I=1, 2, ..., M
C               B)  XX, YY ARE LM BY 1 ARRAYS TO HOLD COORDINATES OF NODES
C               C)  LINE, LL, II, NV, NTR HAVE BEEN EXPLAINED ABOVE
C               D)  GAMMA AND ALPHA ARE CLEAR
C
C           8. NOTE THAT A, B, C, XX, YY, I, I1, I2, J1, J2, J3, DELTA ARE LABELLED
C               COMMON STORAGE SO THAT IN ANY SUBPROGRAM WE KNOW THAT
C               TRIANGLE I HAS VERTICES (XX(J1), YY(J1)), (XX(J2), YY(J2)),
C               (XX(J3), YY(J3)). THAT IS, VERTEX 1 IS NODE J1, VERTEX 2 IS
C               NODE J2, VERTEX 3 IS NODE J3 UNLESS A LINE INTEGRAL IS
C               INVOLVED IN WHICH CASE THE LINE INTEGRAL GOES FROM NODE J1
C               TO NODE J2 IN TRIANGLE I OR UNLESS VERTEX I1 IS NODE J1
C               AND VERTEX I2 IS NODE J2 - THE BASIS FUNCTIONS INVOLVE
C               A(I, I1)+B(I, I1)*X+C(I, I1)**Y FOR VERTEX I1 IN TRIANGLE I
C               AND A(I, I2)+B(I, I2)*X+C(I, I2)*Y FOR VERTEX I2 IN TRIANGLE I
C               DELTA IS 1/2 THE AREA OF TRIANGLE I
C
C       TO CHANGE PROBLEMS:
C           1) CHANGE FUNCTION SUBPROGRAMS P, Q, R, F, G, G1, G2
C           2) CHANGE DATA INPUT FOR K, N, M, LN, LM, NL.
C           3) CHANGE DATA INPUT FOR XX, YY, LLL, II, NV.
C           4) CHANGE DATA INPUT FOR LINE.
C           5) CHANGE DIMENSION STATEMENTS TO READ :
C               DIMENSION A(M, 3), B(M, 3), C(M, 3), XX(LM), YY(LM)
C                       THERE ARE 10 PLACES TO CHANGE.
C               DIMENSION LINE(NL, 2), LL(LM), II(LM, MAX LL(LM)),
C                   NV(LM, MAX LL(LM))—1 PLACE TO CHANGE
C               DIMENSION NTR(M, 3), GAMMA(LM), ALPHA(LN, LN+1)
C                       THERE IS 1 PLACE TO CHANGE
C
        EXTERNAL P, Q, RR, FFF, GG1, GG2, GG3, GG4, GG5
        COMMON A(10, 3), B(10, 3), C(10, 3), XX(11), YY(11), I, I1, I2, J1, J2, J3
        COMMON DELTA
        DIMENSION LINE(5, 2), LL(11), II(11, 5), NV(11, 5)
```

277

```
         DIMENSION NTR(10,3),GAMMA(11),ALPHA(5,6)
C        INPUT K,N,M,LN,LM,NL
         READ(5,1) K,N,M,LN,LM,NL
1        FORMAT(6I2)
C        INPUT COORDINATES OF EACH NODE, NUMBER OF TRIANGLES NODE IS IN
C        (LLL) , WHICH TRIANGLES THE NODE IS IN, AND WHICH VERTEX OF
C        EACH TRIANGLE THE NODE IS IN
         DO 2 KK=1,LM
             READ(5,3) XX(KK),YY(KK),LLL,(II(KK,J),NV(KK,J),J=1,LLL)
3            FORMAT(2E15.8,I2,5(I2,I2))
             LL(KK)=LLL
C            COMPUTE ENTRIES FOR NTR
             DO 4 J=1,LLL
                 N1=II(KK,J)
                 N2=NV(KK,J)
4                NTR(N1,N2)=KK
2        CONTINUE
C        INPUT LINE INFORMATION
         IF(NL.GT.0) THEN DO
             READ(5,10) ((LINE(I,J),J=1,2),I=1,NL)
10           FORMAT(10I2)
         END IF
         K1=K+1
         N1=LN+1
C        OUTPUT INFORMATION ON TRIANGLES, VERTICES AND NODES
         WRITE(6,100)
100      FORMAT(1H0,54X,'TRIANGLES ARE AS FOLLOWS')
         WRITE(6,101)
101      FORMAT(1H0,20X,'TRIANGLE',20X,'VERTEX 1',20X,'VERTEX 2',20X,
        1'VERTEX 3',/)
         WRITE(6,102) (I,(NTR(I,J),J=1,3),I=1,M)
102      FORMAT(24X,I3,21X,'NODE ',I3,20X,'NODE ',I3,20X,'NODE ',I3)
         WRITE(6.103)
103      FORMAT(//,25X,'NODE',25X,'X-COORD.',25X,'Y-COORD.',/)
         WRITE(6,104) (KK,XX(KK),YY(KK),KK=1,LM)
104      FORMAT(26X,I3,22X,E15.8,18X,E15.8)
C        OUTPUT INFO ON LINES
         WRITE(6,109)
109      FORMAT(////,45X,'DESCRIPTION OF LINES FOR LINE INTEGRALS',//)
         WRITE(6,110) (L1,LINE(L1,1),LINE(L1,2),L1=1,NL)
110      FORMAT(45X,'LINE',I3,'  FROM NODE',I3,' TO NODE',I3)
C        STEP 1
         IF(LM.NE.LN) THEN DO
             DO 5 L=N1,LM
5            GAMMA(L)=G(XX(L),YY(L))
         END IF
C        STEP 2 - INITIALIZATION OF ALPHA AND NOTE THAT
C        ALPHA(I,LN+1)=BETA(I)
         DO 6 I=1,LN
```

```
              DO 6 J=1,N1
6             ALPHA(I,J)=0.0
C     STEPS 3,4 AND 6 THROUGH 12 ARE WITHIN THE NEXT LOOP
C     FOR EACH TRIANGLE I LET NODE J1 BE VERTEX 1,NODE J2 BE VERTEX 2,
C     AND NODE J3 BE VERTEX 3
C     STEP 3
      DO 7 I=1,M
          J1=NTR(I,1)
          J2=NTR(I,2)
          J3=NTR(I,3)
          DELTA=XX(J2)*YY(J3)-XX(J3)*YY(J2)-XX(J1)*(YY(J3)-YY(J2))
     *    +YY(J1)*(XX(J3)-XX(J2))
          A(I,1)=(XX(J2)*YY(J3)-YY(J2)*XX(J3))/DELTA
          B(I,1)=(YY(J2)-YY(J3))/DELTA
          C(I,1)=(XX(J3)-XX(J2))/DELTA
          A(I,2)=(XX(J3)*YY(J1)-YY(J3)*XX(J1))/DELTA
          B(I,2)=(YY(J3)-YY(J1))/DELTA
          C(I,2)=(XX(J1)-XX(J3))/DELTA
          A(I,3)=(XX(J1)*YY(J2)-YY(J1)*XX(J2))/DELTA
          B(I,3)=(YY(J1)-YY(J2))/DELTA
          C(I,3)=(XX(J2)-XX(J1))/DELTA
C         STEP 4
C         I1=J FOR STEP 4 AND I1=K FOR STEP 7
          DO 8 I1=1,3
C         STEP 8
              JJ1=NTR(I,I1)
C             I2=K FOR STEP 4 AND I2=J FOR STEP 9
              DO 9 I2=1,I1
C                 STEP 10 AND STEP 4
                  JJ2=NTR(I,I2)
C                 ZZ=Z(I1,I2)**I
                  ZZ=B(I,I1)*B(I,I2)*QQ(P)+C(I,I1)*C(I,I2)*QQ(Q)
     *                -QQ(RR)
C                 STEP 11 AND 12
                  IF(JJ1.LE.LN) THEN DO
                      IF(JJ2.LE.LN) THEN DO
                          ALPHA(JJ1,JJ2)=ALPHA(JJ1,JJ2) + ZZ
                          IF(JJ1.NE.JJ2) ALPHA(JJ2,JJ1) =
     *                        ALPHA(JJ2,JJ1)+ZZ
                      ELSE DO
                          ALPHA(JJ1,N1)=ALPHA(JJ1,N1)-ZZ*GAMMA(JJ2)
                      END IF
                  ELSE DO
                      IF(JJ2.LE.LN) ALPHA(JJ2,N1)=ALPHA(JJ2,N1)
     *                    -ZZ*GAMMA(JJ1)
                  END IF
9             CONTINUE
          HH=-QQ(FFF)
          IF(JJ1.LE.LN) ALPHA(JJ1,N1)=ALPHA(JJ1,N1)+HH
8         CONTINUE
```

```
7         CONTINUE
C         OUTPUT BASIS FUNCTIONS
          WRITE(6, 105)
105       FORMAT(////, 41X, 'BASIS FUNCTIONS ON EACH TRIANGLE')
          WRITE(6, 106)
106       FORMAT(1H0, 8X, 'TRIANGLE', 6X, 'VERTEX', 9X, 'NODE', 33X, 'FUNCTION', /)
          DO 108 I=1, M
          DO 108 J=1, 3
108           WRITE(6, 107) I, J, NTR(I, J), A(I, J), B(I, J), C(I, J)
107           FORMAT(10X, I3, 10X, I3, 10X, I3, 10X, E15. 8, ' + ', E15. 8, ' * X + '
      *       , E15. 8, ' * Y')
C     STEP 5
C     FOR EACH LINE SEGMENT JI=1,...,NL AND FOR EACH TRIANGLE I,
C     I=K1,...,N WHICH MAY CONTAIN LINE JI. SEARCH ALL 3 VERTICES
C     FOR POSSIBLE CORRESPONDENCE
C     STEP 5 AND STEPS 13 THROUGH 19
      IF(NL. NE. 0. AND. N. NE. K) THEN DO
              DO 11 JI=1, NL
                  DO 12 I=K1, N
                      DO 13 I1=1, 3
C                     I1=J IN STEP 5 AND I1=K IN STEP 14
C                     STEP 15
                      J1=NTR(I, I1)
                      IF(LINE(JI, 1). EQ. J1) THEN DO
                          DO 14 I2=1, 3
C                         I2=K IN STEP 5 AND I2=J IN STEP 16
C                         STEP 17
                          J2=NTR(I, I2)
                          IF(LINE(JI, 2). EQ. J2) THEN DO
C                         WE HAVE CORRESPONDENCE OF VERTEX I1
C                         IN TRIANGLE I WITH NODE J1 AS START OF LINE JI
C                         AND VERTEX I2 WITH NODE J2 AS END OF LINE JI
C                             STEP 5
                              XJ=SQ(GG1)
                              XJ1=SQ(GG4)
                              XJ2=SQ(GG5)
                              XI1=SQ(GG2)
                              XI2=SQ(GG3)
C                             STEP 8 AND 19
                              IF(J1. LE. LN) THEN DO
                                  IF(J2. LE. LN) THEN DO
                                      ALPHA(J1, J1)=ALPHA(J1, J1)+XJ1
                                      ALPHA(J1, J2)=ALPHA(J1, J2)+XJ
                                      ALPHA(J2, J2)=ALPHA(J2, J2)+XJ2
                                      ALPHA(J2, J1)=ALPHA(J2, J1)+XJ
                                      ALPHA(J1, N1)=ALPHA(J1, N1)+XI1
                                      ALPHA(J2, N1)=ALPHA(J2, N1)+XI2
                                  ELSE DO
                                      ALPHA(J1, N1)=ALPHA(J1, N1)-
```

280

```
     *                       XJ*GAMMA(J2)+XI1
                             ALPHA(J1, J1)=ALPHA(J1, J1)+XJ1
                          END IF
                       ELSE DO
                          IF(J2. LE. LN) THEN DO
                             ALPHA(J2, N1)=ALPHA(J2, N1)-
     *                          XJ*GAMMA(J1)+XI2
                             ALPHA(J2, J2)=ALPHA(J2, J2)+XJ2
                          END IF
                       END IF
                    END IF
14                  CONTINUE
                 END IF
13               CONTINUE
12            CONTINUE
11         CONTINUE
        END IF
C       OUTPUT ALPHA
        WRITE(6, 111)
111     FORMAT(////, 50X, 'ENTRIES OF ALPHA', /)
        WRITE(6, 112) ((I, J, ALPHA(I, J), J=1, N1), I=1, LN)
112     FORMAT(3(3X, 'ALPHA(', I3, ',', I3, ') = ', E15. 8))
C       STEP 20
        IF(LN. GT. 1) THEN DO
        INN=LN-1
        DO 30 I=1, INN
           I1=I+1
           DO 31 J=I1, LN
              CCC=ALPHA(J, I)/ALPHA(I, I)
              DO 32 J1=I1, N1
32            ALPHA(J, J1)=ALPHA(J, J1)-CCC*ALPHA(I, J1)
31            ALPHA(J, I)=0. 0
30      CONTINUE
        END IF
        GAMMA(LN)=ALPHA(LN, N1)/ALPHA(LN, LN)
        IF(LN. GT. 1) THEN DO
        DO 33 I=1, INN
           J=LN-I
           CCC=ALPHA(J, N1)
           J1=J+1
           DO 34 KK=J1, LN
34            CCC=CCC-ALPHA(J, KK)*GAMMA(KK)
33         GAMMA(J)=CCC/ALPHA(J, J)
        END IF
C       STEP 21
C       OUTPUT GAMMA
        WRITE(6, 113)
113     FORMAT(//, 56X, 'COEFFICIENTS GAMMA', /)
        DO 115 I=1, LM
```

```
         LLL=LL(I)
115      WRITE(6,114) I,GAMMA(I),I,(II(I,J),J=1,LLL)
114      FORMAT(23X,'GAMMA(',I3,') = ',E15.8,'  USED WITH NODE',I3,' IN TRI
        1ANGLES ',5(I3,2X))
C        STEP 23
         STOP
         END
         FUNCTION P(X,Y)
         P=1
         RETURN
         END
         FUNCTION Q(X,Y)
         Q=1
         RETURN
         END
         FUNCTION R(X,Y)
         R=0
         RETURN
         END
         FUNCTION F(X,Y)
         F=0
         RETURN
         END
         FUNCTION G(X,Y)
         G=4
         RETURN
         END
         FUNCTION G1(X,Y)
         G1=0.0
         RETURN
         END
         FUNCTION G2(X,Y)
         G2=0
         T=1.0E-05
         IF((.2-T.LE.X.AND.X.LE.4+T).AND.ABS(Y-.2).LE.T) G2=X
         IF((.5-T.LE.X.AND.X.LE..6+T).AND.ABS(Y-.1).LE.T) G2=X
         IF((-T.LE.Y.AND.Y.LE..1+T).AND.ABS(X-.6).LE.T) G2=Y
         IF((-T.LE.X.AND.X.LE..2+T).AND.ABS(Y+X-.4).LE.T)G2=(X+Y)/SQRT(
        * 2.)
         IF((.4-T.LE.X.AND.X.LE..5+T).AND.ABS(Y+X-.6).LE.T)G2=(X+Y)/SQRT(
        * 2.)
         RETURN
         END
         FUNCTION RR(X,Y)
         COMMON A(10,3),B(10,3),C(10,3),XX(11),YY(11),I,I1,I2,J1,J2,J3
         COMMON DELTA
         RR=R(X,Y)*(A(I,I1)+B(I,I1)*X+C(I,I1)*Y)*(A(I,I2)+B(I,I2)*X+C(I,I2)
        1*Y)
         RETURN
```

```
      END
      FUNCTION FFF(X, Y)
      COMMON A(10, 3), B(10, 3), C(10, 3), XX(11), YY(11), I, I1, I2, J1, J2, J3
      COMMON DELTA
      FFF=F(X, Y)*(A(I, I1)+B(I, I1)*X+C(I, I1)*Y)
      RETURN
      END
      FUNCTION GG1(X, Y)
      COMMON A(10, 3), B(10, 3), C(10, 3), XX(11), YY(11), I, I1, I2, J1, J2, J3
      COMMON DELTA
      GG1=G1(X, Y)*(A(I, I1)+B(I, I1)*X+C(I, I1)*Y)*(A(I, I2)+B(I, I2)*X+
     1C(I, I2)*Y)
      RETURN
      END
      FUNCTION GG2(X, Y)
      COMMON A(10, 3), B(10, 3), C(10, 3), XX(11), YY(11), I, I1, I2, J1, J2, J3
      COMMON DELTA
      GG2=G2(X, Y)*(A(I, I1)+B(I, I1)*X+C(I, I1)*Y)
      RETURN
      END
      FUNCTION GG3(X, Y)
      COMMON A(10, 3), B(10, 3), C(10, 3), XX(11), YY(11), I, I1, I2, J1, J2, J3
      COMMON DELTA
      GG3=G2(X, Y)*(A(I, I2)+B(I, I2)*X+C(I, I2)*Y)
      RETURN
      END
      FUNCTION GG4(X, Y)
      COMMON A(10, 3), B(10, 3), C(10, 3), XX(11), YY(11), I, I1, I2, J1, J2, J3
      COMMON DELTA
      GG4=G1(X, Y)*(A(I, I1)+B(I, I1)*X+C(I, I1)*Y)**2
      RETURN
      END
      FUNCTION GG5(X, Y)
      COMMON A(10, 3), B(10, 3), C(10, 3), XX(11), YY(11), I, I1, I2, J1, J2, J3
      COMMON DELTA
      GG5=G1(X, Y)*(A(I, I2)+B(I, I2)*X+C(I, I2)*Y)**2
      RETURN
      END
      FUNCTION QQ(FF)
      COMMON A(10, 3), B(10, 3), C(10, 3), XX(11), YY(11), I, I1, I2, J1, J2, J3
      COMMON DELTA
      A1=(XX(J1)+XX(J2)+XX(J3))/3
      A2=(YY(J1)+YY(J2)+YY(J3))/3
      QQ=3*(FF(XX(J1), YY(J1))+FF(XX(J2), YY(J2))+FF(XX(J3), YY(J3)))/60
     1+8*(FF((XX(J1)+XX(J2))/2, (YY(J1)+YY(J2))/2)+FF((XX(J1)+XX(J3))/2,
     1(YY(J1)+YY(J3))/2)+FF((XX(J2)+XX(J3))/2, (YY(J2)+YY(J3))/2))/60
     1+27*FF(A1, A2)/60
      QQ=QQ*ABS(DELTA)/2
      RETURN
```

```
          END
          FUNCTION SQ(PP)
          COMMON A(10,3),B(10,3),C(10,3),XX(11),YY(11),I,I1,I2,J1,J2,J3
          COMMON DELTA
          DIMENSION X(101)
          F1(T)=PP(T1*T+X1,T2*T+Y1)*SQRT(T1*T1+T2*T2)
          X1=XX(J1)
          Y1=YY(J1)
          X2=XX(J2)
          Y2=YY(J2)
          T1=X2-X1
          T2=Y2-Y1
          H=.01
          DO 2 II=1,101
2         X(II)=(II-1)*H
          SQ=(F1(X(1))+F1(X(101)))*H/3
          Q1=0.0
          Q2=0.0
          DO 3 II=1,49
          Q1=Q1+2*H*F1(X(2*II+1))/3
3         Q2=Q2+4*H*F1(X(2*II))/3
          Q2=Q2+4*H*F1(X(100))/3
          SQ=SQ+Q1+Q2
          RETURN
          END
$ENTRY
000410051105
.2E+00          .2E+00          0501030201050306030701
.4E+00          .2E+00          03020303010903
.3E+00          .1E+00          0402020703080301
.5E+00          .1E+00          03030304011003
.6E+00          .1E+00          010403
.0E+00          .4E+00          010101
.0E+00          .2E+00          0201020501
.0E+00          .0E+00          0205020601
.2E+00          .0E+00          03060207020801
.4E+00          .0E+00          040302080209021001
.6E+00          .0E+00          0204021002
0106020104020504115
/*
```

284